Multimodal Learning toward Micro-Video Understanding

Synthesis Lectures on Image, Video, and Multimedia Processing

Editor
Alan C. Bovik, *University of Texas, Austin*

The Lectures on Image, Video and Multimedia Processing are intended to provide a unique and groundbreaking forum for the world's experts in the field to express their knowledge in unique and effective ways. It is our intention that the Series will contain Lectures of basic, intermediate, and advanced material depending on the topical matter and the authors' level of discourse. It is also intended that these Lectures depart from the usual dry textbook format and instead give the author the opportunity to speak more directly to the reader, and to unfold the subject matter from a more personal point of view. The success of this candid approach to technical writing will rest on our selection of exceptionally distinguished authors, who have been chosen for their noteworthy leadership in developing new ideas in image, video, and multimedia processing research, development, and education.

In terms of the subject matter for the series, there are few limitations that we will impose other than the Lectures be related to aspects of the imaging sciences that are relevant to furthering our understanding of the processes by which images, videos, and multimedia signals are formed, processed for various tasks, and perceived by human viewers. These categories are naturally quite broad, for two reasons: First, measuring, processing, and understanding perceptual signals involves broad categories of scientific inquiry, including optics, surface physics, visual psychophysics and neurophysiology, information theory, computer graphics, display and printing technology, artificial intelligence, neural networks, harmonic analysis, and so on. Secondly, the domain of application of these methods is limited only by the number of branches of science, engineering, and industry that utilize audio, visual, and other perceptual signals to convey information. We anticipate that the Lectures in this series will dramatically influence future thought on these subjects as the Twenty-First Century unfolds.

Multimodal Learning toward Micro-Video Understanding

Liqiang Nie, Meng Liu, and Xuemeng Song

ISBN: 978-3-031-01127-6   paperback
ISBN: 978-3-031-02255-5   ebook
ISBN: 978-3-031-00216-8   hardcover

DOI 10.1007/978-3-031-02255-5

A Publication in the Springer series
SYNTHESIS LECTURES ON IMAGE, VIDEO, AND MULTIMEDIA PROCESSING

Lecture #20
Series Editor: Alan C. Bovik, *University of Texas, Austin*
Series ISSN
Print 1559-8136   Electronic 1559-8144

Multimodal Learning toward Micro-Video Understanding

Liqiang Nie, Meng Liu, and Xuemeng Song
Shandong University, Jinan, China

SYNTHESIS LECTURES ON IMAGE, VIDEO, AND MULTIMEDIA PROCESSING #20

ABSTRACT

Micro-videos, a new form of user-generated contents, have been spreading widely across various social platforms, such as Vine, Kuaishou, and TikTok. Different from traditional long videos, micro-videos are usually recorded by smart mobile devices at any place within a few seconds. Due to its brevity and low bandwidth cost, micro-videos are gaining increasing user enthusiasm. The blossoming of micro-videos opens the door to the possibility of many promising applications, ranging from network content caching to online advertising. Thus, it is highly desirable to develop an effective scheme for the high-order micro-video understanding.

Micro-video understanding is, however, non-trivial due to the following challenges: (1) how to represent micro-videos that only convey one or few high-level themes or concepts; (2) how to utilize the hierarchical structure of the venue categories to guide the micro-video analysis; (3) how to alleviate the influence of low-quality caused by complex surrounding environments and the camera shake; (4) how to model the multimodal sequential data, i.e., textual, acoustic, visual, and social modalities, to enhance the micro-video understanding; and (5) how to construct large-scale benchmark datasets for the analysis? These challenges have been largely unexplored to date.

In this book, we focus on addressing the challenges presented above by proposing some state-of-the-art multimodal learning theories. To demonstrate the effectiveness of these models, we apply them to three practical tasks of micro-video understanding: popularity prediction, venue category estimation, and micro-video routing. Particularly, we first build three large-scale real-world micro-video datasets for these practical tasks. We then present a multimodal transductive learning framework for micro-video popularity prediction. Furthermore, we introduce several multimodal cooperative learning approaches and a multimodal transfer learning scheme for micro-video venue category estimation. Meanwhile, we develop a multimodal sequential learning approach for micro-video recommendation. Finally, we conclude the book and figure out the future research directions in multimodal learning toward micro-video understanding.

KEYWORDS

micro-video understanding, multimodal transductive learning, multimodal cooperative learning, multimodal transfer learning, multimodal sequential learning, popularity prediction, venue category estimation, micro-video recommendation

Contents

Preface

The unprecedented growth of portable devices contributes to the success of micro-video sharing platforms such as Vine, Kuaishou, and TikTok. These devices enable users to record and share their daily life within a few seconds in the form of micro-videos at any time and any place. As a new media type, micro-videos have garnered great enthusiasm due to brevity, authenticity, communicability, and low-cost. The proliferation of micro-videos confirms the old saying that good things come in small packages.

Like traditional long videos, micro-videos are a combination of textual, acoustic, and visual modalities. These modalities are correlated rather than independent, and they essentially characterize the same micro-videos from distinct angles. Effectively fusing heterogeneous modalitics toward video understanding indeed has been well-studied in the past decade. Yet, micro-videos have their unique characteristics and corresponding research challenges, including but not limited to the following.

(1) Information sparseness. Micro-videos are very short, lasting for 6–15 s, and they hence usually convey only a few concepts. In light of this, we need to learn their sparse and conceptual representations for better discrimination. (2) Hierarchical structure. Micro-videos are implicitly organized into a four-layer hierarchical tree structure with respect to their recording venues. We should leverage such a structure to guide the organization of micro-videos by categorizing them into the leaf nodes of this tree. (3) Low-quality. Most portable devices have nothing to offer for video stabilization. Some recorded videos can thus be visually shaky or bumpy, which greatly hinders the visual expression. Furthermore, the audio track that comes along with the video can differ in terms of distortion and noise, such as buzzing, hums, hisses, and whistling, which is probably caused by the poor microphones or complex surrounding environments. We thus have to harness the external visual or sound knowledge to compensate the shortest boards. (4) Multimodal sequential data. Beyond textual, acoustic, and visual modalities, micro-videos also have social modality. In such a context, a user is enabled to interact with micro-videos and other users via social actions , such as click, like, and follow. As time goes on, multiple sequential data in different forms emerge and reflect users' historical preferences. To strengthen micro-video understanding, we have to characterize and model the sequential patterns. (5) The last challenge we are facing is the lack of benchmark datasets to justify our ideas.

In this book, to tackle the aforementioned research challenges, we present some state-of-the-art multimodal learning theories and verify them over three practical tasks of micro-video understanding: popularity prediction, venue category estimation, and micro-video routing. In particular, we first construct three large-scale real-world micro-video datasets corresponding to the three practical tasks. We then propose a multimodal transductive learning framework to learn the micro-video representations in an optimal latent space via unifying and preserving information from different modalities. In this transductive framework, we integrate the low-rank constraints to somehow alleviate the information sparseness and low-quality problems. This framework is verified on the popularity prediction task. We next present a series of multimodal cooperative learning approaches, which explicitly model the consistent and complementary modality correlations. In the multimodal cooperative learning approaches, we make full use of the hierarchical structure by the tree-guided group lasso, and further solve the information sparseness via dictionary learning. Following that, we work toward compensating the low-quality acoustic modalities via harnessing the external sound knowledge. This is accomplished by a deep multimodal transfer learning scheme. The multimodal cooperative learning approaches and the multimodal transfer learning scheme are both justified over the task of venue category estimation. Thereafter, we develop a multimodal sequential learning approach, relying on temporal graph-based long short-term memory networks, to intelligently route micro-videos to the target users in a personalized manner. We ultimately summarize the book and figure out the future research directions in multimodal learning toward micro-video understanding.

This book represents a preliminary research on learning from multiple correlated modalities of given micro-videos, and we anticipate that the lectures in this series will dramatically influence future thought on these subjects. If in this book we have been able to dream further than others have, it is because we are standing on the shoulders of giants.

Liqiang Nie, Meng Liu, and Xuemeng Song
July 2019

Acknowledgments

It is a pleasure to acknowledge many colleagues who have made this time-consuming book project possible and enjoyable. In particular, many members of the iLearn Center in Shandong University and the LMS Lab in National University of Singapore have co-worked on various aspects of multimodal learning and its applications in micro-video understanding. Their efforts have supplied ingredients for insightful discussions related to the writing of this book, and hence we are greatly appreciative.

Our first thanks undoubtedly goes to Dr. Peiguang Jing at Tianjin University, Dr. Jingyuan Chen at Alibaba Group, Dr. Jianglong Zhang at Information & Telecommunication Company SGCC, as well as Mr. Yongqi Li and Mr. Yinwei Wei at Shandong University. We consulted with them on some specific technical chapters of the book and they are also the major contributors of some chapters. Their constructive feedback and comments at various stages have been significantly helpful in shaping the book. We also take this opportunity to thank Prof. Tat-Seng Chua at National University of Singapore who never hesitated to offer his advice and share his valuable experience whenever the authors needed him. Particular thanks go to Miss Qian Liu and Miss Xiaoli Li, who read the earlier drafts of the manuscript and provided helpful comments to improve the readability.

We are very grateful to the anonymous reviewers. Despite their busy schedules, they read the book very carefully and gave us many insightful suggestions, which were the key to making this book as sound as possible.

We are grateful to Morgan & Claypool and particularly the Vice President & Publisher Mr. Joel Claypool for his help and patience throughout the writing of this book. He has managed to get everything done on time and provided us with many pieces of valuable advice. This book would not have been completed, or at least not be what it looks like now, without the support, direction, and help of him and his team.

Last, but certainly no least, our thanks go to our beloved families for their unwavering support during this fun book project, as well as for their understanding and tolerance of many weekends and long nights spent on the book by the authors. We dedicate this book to them, with love.

Liqiang Nie, Meng Liu, and Xuemeng Song
July 2019

CHAPTER 1

Introduction

1.1 MICRO-VIDEO PROLIFERATION

Traditional long video sharing platforms, like Youtube[1] and Youku,[2] were very successful before 2010. They encourage professionals to capture and upload high-quality long videos online, and while also allowing ordinary users to consume long video contents. After 2010, however, we are indeed living in a new era of ever-dwindling attention span and are hungry for the quick content, e.g., tweets and micro-blogs instead of the complicated documents. In such a climate, the way that the Internet users digest videos also had a dramatic shift. Not unnaturally, the bite-sized videos embracing the philosophy of "shorter is better," are becoming popular with the rise of some services, like Vine,[3] Snapchat,[4] Viddy,[5] and MixBit,[6] which limit their video lengths up to 6, 10, 15, and 16 s, respectively. As a result, we have dubbed such short videos "micro-videos." In Table 1.1, we summarize some mainstream micro-video services and their statistical data.

Formally, micro-videos are defined as videos that are usually captured by ordinary users via mobile devices to record a few seconds of their daily lives to share over social media platforms. The properties of easy-to-operate, instant sharing, down-to-the-earth content make micro-video services unexpectedly popular, especially among the grassroots. Considering Kuaishou as an example, as of January 2019, it had reached 190 million active users and 10 million uploaded micro-videos daily, and each active online user usually spends up to 1 h to share, search, and view their interested micro-videos on Kuaishou. Figure 1.1 demonstrates one Kuaishou micro-video example.

As micro-videos have surged in recent years, micro-video platforms have become successful online communities that penetrate many aspects of society. First, it largely enriches people's life by providing valuable chances for people to share their life, encounter different people, and make friends. Second, it facilitates the spread of traditional culture. In the TikTok platform, an app owning more than 500 million active users globally as of June 2018, the top culture categories are calligraphy and painting, traditional crafts, and opera. For one opera-related competition,

[1]https://www.youtube.com/
[2]https://www.youku.com/
[3]https://vine.com/
[4]https://www.snapchat.com/
[5]http://www.fullscreen.com/
[6]https://mixbit.com/home

Table 1.1: Examples of the current popular micro-video platforms

Platforms	Kuaishou[7]	Huoshan[8]	Instagram[9]	Snapchat	TikTok[10]
Monthly Active Users (million)	321 (2019.1)	47 (2017.9)	330 (2018.8)	300 (2018.9)	500 (2018.6)
Duration (s)	11 or 57	15	3–60	10	15
Date of Establishment	2011.7	2016.4	2010.10	2009.4	2017.5
Minutes Users Spend Online on Average	60 (2019.1)	59 (2017.9)	53 (2018.8)	40 (2018.9)	52 (2018.6)

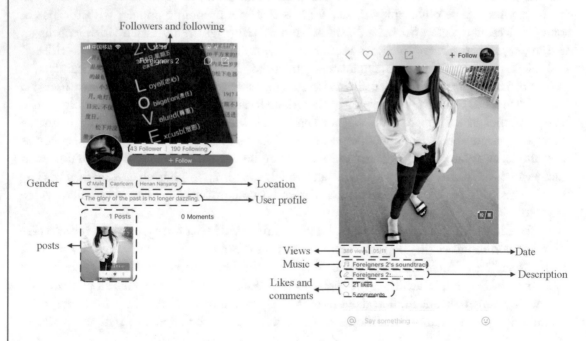

Figure 1.1: Illustration of a micro-video example selected from the Kuaishou App.

[7]www.kuaishou.com
[8]www.huoshanzhibo.com
[9]www.instagram.com
[10]www.tiktok.com

180,000 people attended with 93% were the young. Third, it delivers an opportunity for remote counties to advertise their rustic scenery. Daocheng county,[11] a rural town in south-west China, gained over 10 million likes on TikTok in 2018. In summary, micro-videos are rocking and taking over the content and social media marketing spaces [120].

Micro-video arises as a new form of user-generated content (UCG) and hence the related research is relatively sparse. The first work [136] on micro-videos is about the creativity assessment. Redi et al. studied the audio-visual features of creative and non-creative videos. Meanwhile, they introduced a computational framework to automatically predict these categories. In the same year, Sano et al. [142] observed that the loop is one of the key features of popular micro-videos, but there are so many non-loop videos mistakenly tagged with "loop." Inspired by this observation, they proposed a degree-of-loop method to measure micro-videos. In this book, we comprehensively work toward analyzing the unique characteristics of micro-videos, designing theoretical solutions for micro-video understanding, and verifying them over several practical tasks.

1.2   PRACTICAL TASKS

In this book, we introduce three practical tasks of micro-video understanding, namely popularity prediction, venue category estimation, and micro-video routing. These three tasks are leveraged to verify several state-of-the-art multimodal learning theories we proposed in this book.

1.2.1   MICRO-VIDEO POPULARITY PREDICTION

According to our observation, among the tremendous volume of micro-videos, some popular ones will be widely viewed and spread by users, while many only gain little attention. This phenomena is similar to many existing social media sites, such as Twitter.[12] For example, one micro-video about the explosion that interrupted during the France-Germany soccer match in 2015 has been successfully looped by over 330 million times. Obviously, if we can identify the hot and popular micro-videos in advance, it will benefit many applications, such as online marketing and network reservation. Regarding online marketing, the accurate early prediction of popular micro-videos can facilitate companies planning advertising campaigns and thus maximizing their revenues. For network service providers, they can timely reserve adequate distributed storage and bandwidth for popular ones, based on the prediction. Therefore, it is highly desirable to develop an effective scheme to accurately predict the popularity of micro-videos.

1.2.2   MICRO-VIDEO VENUE CATEGORIZATION

Organizing micro-videos plays an increasingly pivotal role in high-order analysis of micro-videos, such as search, browse, and navigation. However, micro-videos are somehow unor-

---

[11]https://en.wikipedia.org/wiki/Daocheng_County
[12]https://twitter.com/

ganized. Different from the traditional long videos that can be well-structured into a specific video genre from "Crime," "Documentary," "Romance," to "War," like the video organization in YouTube, micro-video, as a new medium, does not have a matured taxonomy to follow. Also, it is inappropriate to apply the long video taxonomy to micro-videos due to their different emphases. In particular, micro-videos record the actual things in life, whereas the long ones cover a broader range of things like marvelous performance.

Thankfully, a micro-video is frequently recorded at a specific place within one shot, and hence micro-video services are able to encourage users to manually label the micro-videos with GPS-suggested venue information [49], such as "Orchard ION Singapore." Each venue belongs to a venue category, such as "shopping mall," based on the Foursquare API[13] and the venue categories are organized into a tree-like taxonomy.[14] We show part of the tree structure in the Figure 1.2. From the figure, we find that: (1) Foursquare organizes the venue categories into a four-layer hierarchical tree structure; and (2) the top layer of the tree contains ten non-leaf nodes (coarse venue categories). We aim to organize micro-videos by categorizing them into the leaf nodes of this tree. It will benefit multifaceted aspects: (1) footprint recordings—it facilitates users to vividly archive where they were and what they did; (2) personalized applications— such people-centric location data enables precise personalized services, such as suggesting local restaurants, alerting regional weather, and spreading business information to nearby customers; and (3) other location-based services. Location information is helpful for the inference of users' interests, the improvement of activity prediction, and the simplification of landmark-oriented video search.

Despite its significance, users of micro-video platforms have been slow to adopt this geospatial feature: in a random sample over 2 million Vine videos, we found that only 1.22% of the videos are associated with venue information. It is thus highly desirable to infer the missing geographic cues.

1.2.3   MICRO-VIDEO ROUTING

Along with the popularity of micro-videos, users are frequently overwhelmed by their uninterested ones. It becomes increasingly difficult and expensive for users to locate their desired micro-videos from the vast candidates. This is due to the following reasons: (1) existing recommendation systems developed for various communities cannot be straightforwardly applied to route micro-videos, since users in micro-video platforms have their unique characteristics, such as the complex interactions between users and micro-videos; and (2) in micro-video platforms, users follow their interested topics and the platforms usually recommend users with the micro-videos falling in the range of the followed topics. However, users' interests are dynamic and evolve over time. In light of this, it is crucial to build a personalized recommendation system to intelligently route micro-videos to the target users, which will strengthen customer stickiness.

---

[13]https://github.com/mLewisLogic/foursquare
[14]https://developer.foursquare.com/categorytree

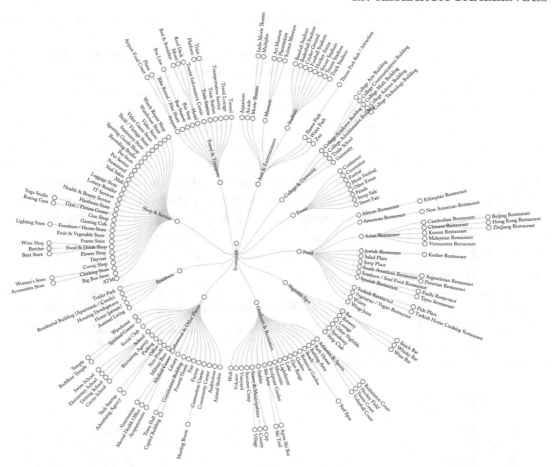

Figure 1.2: The hierarchical structure of the venue categories. We only illustrate part of the structure due to the limited space. The entire tree can be viewed here: `https://developer.fo ursquare.com/categorytree`.

1.3 RESEARCH CHALLENGES

Like traditional long videos, micro-videos are also the unity of visual frames, surrounding textual descriptions, and audio channels. Hereafter, we name them as visual, textual, and acoustic modalities, respectively. These three modalities consistently or complementarily describe a real-life event from distinct angles in different media forms. In a sense, these modalities are correlated rather than independent and essentially characterize the same micro-videos. Therefore, the major challenge lies on how to effectively fuse heterogeneous clues of given micro-videos from multiple modalities. Although researchers have well exploited this direction in the past decade, and proposed many solutions like the early fusion, late fusion, and subspace learning strategies, multimodal fusion is still an open-ended research problem.

In addition to the common ones, micro-videos also have their unique characteristics and corresponding research challenges. The first such challenge is information sparseness and information loss. Unlike the traditional long videos with rich content, micro-videos are very short, only lasting 6–15 s. As illustrated in Figure 1.3, we plot the video length distribution over 303,000 micro-videos collected from Vine. It is obvious that most of them are less than 7 s. Persuasively, videos with short lengths make video production and broadcasting easy, downloading timely, and play fluent on portable devices. However, in contrast to the traditional long videos, the visual information conveyed by micro-videos is somehow inadequate as they usually only contain one or a few high-level concepts and are unable to provide rich contexts for effective similarity measurement. We thus need to learn their sparse and conceptual representations for better discrimination. In addition, apart from acoustic and visual modalities, micro-videos are, more often than not, uploaded with textual descriptions, which express some useful signals that may be not available in the other two modalities. However, the textual information may be not well correlated with visual and acoustic cues. Moreover, according to our statistics based upon 276,624 Vine videos, more than 11.4% of them do not have such text, probably due to users' casual habits. This serious information missing problem greatly reduces the usability of textual modality. It is necessary to exploit the complementarity between different views to learn the latent interconnected patterns in order to address the incomplete problem.

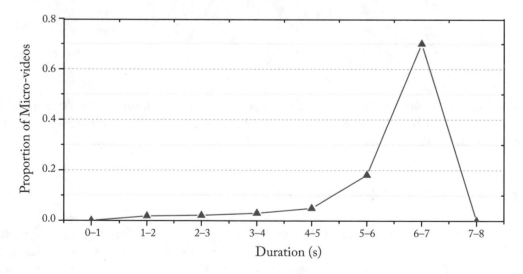

Figure 1.3: Duration distribution over our collected 303,242 micro-videos from Vine.

The second challenge is the hierarchical structure. Micro-videos are implicitly organized into a four-layer hierarchical tree structure with respect to their recording venues. The venue categories in the tree taxonomy are not independent, but hierarchically correlated (as shown in Figure 1.2). In particular, the closer two venue categories are located in the tree, the more similar concepts the associated micro-videos should convey. In a sense, we have to consider the inherent structure of micro-videos when learning their representations. Also, we should leverage such a structure to guide the organization of micro-videos by categorizing them into the leaf nodes of this tree.

The third research challenge we face is low-quality. Micro-videos are often captured by users with hand-held mobile devices. Most portable devices have nothing to offer for video stabilization, which easily results in poor video quality, such as low resolution, wobbly frames, constrained lighting conditions, and background noise. This greatly hinders the visual expression. Furthermore, the audio track that comes along with the video can be in different types of distortion and noise, such as buzzing, hums, hisses, and whistling, which are probably caused by the poor microphones or complex surrounding environments. We thus have to harness the external high-quality visual or sound knowledge to compensate the internal shortest boards. Alternatively, we should build robust models to explore the intrinsic structure property embedded in data by inferring meaningful features and alleviating the impact of noisy ones.

The fourth challenge is the multimodal sequential data. Beyond textual, acoustic, and visual modalities, micro-videos have a new one; social modality. In such a context, a user is enabled to interact with micro-videos and other users via social actions, such as click, like, and follow. As time goes on, multiple sequential data in different forms emerge and they reflect users' historical preferences. To strengthen micro-video understanding, we have to characterize and model the sequential patterns.

Last, but not least, the final research problem we are facing is the lack of benchmark datasets to support our research on micro-video understanding.

1.4   OUR SOLUTIONS

In this book, to tackle the aforementioned research challenges, we present some state-of-the-art multimodal learning theories and verify them over three practical tasks of micro-video understanding: popularity prediction, venue category estimation, and micro-video routing.

We first construct two large-scale, real-world micro-video datasets, from Vine, which correspond to the tasks of popularity prediction and venue category estimation. Considering the large volume of micro-videos, to save the human labor, we establish their ground truth using some manually defined rules. As to the task of micro-video routing, we carry on our experiments on the public benchmark datatsets.

For the popularity prediction task, we propose a multimodal transductive learning framework to learn the micro-video representations in an optimal latent space. We assume that this optimal latent space maintains the original intrinsic characteristics of micro-videos in the original spaces. In light of this, all modalities are forced to be correlated. Meanwhile, micro-videos with different popularity can be better separated in such optimal common space, as compared to that of each single modality. In this transductive framework, we integrate the low-rank constraints to somehow alleviate the information sparseness and low-quality problems. Because the formulated objective function is not smooth and hard to solve, we design an effective algorithm based on the augmented Lagrange multiplier to optimize it and ensure a fast convergence.

As to the venue category estimation of micro-videos, we shed light on characterizing and modeling the correlations between modalities, especially the consistent and complementary relations. The consistent part is to strengthen the confidence and the complementary one is able to supplement a lot of exclusive information. We argue that explicitly parsing these two kinds of correlations and treating them separately within a unified model can boost the representation discrimination for multimodal samples. Toward this goal, we devise a series of cooperative learning models, which split the consistent information from the complementary one, and leverage them to strengthen the expressiveness of each modality. In addition, we regularize the hierarchical structure of micro-videos via the tree-guided group lasso, which characterizes the inter- and intra-relatedness among venue categories. Meanwhile, we integrate the dictionary learning component to learn the concept-level sparse representation of micro-videos, whereby the problem of information sparseness can be alleviated.

We also explore methods to enhance the low-quality problems for the venue category estimation task, especially compensating the acoustic modality. Toward this goal, we develop a deep multimodal transfer learning scheme, which transfers knowledge from external high-quality sound clips to strengthen the description of the internal acoustic modality in micro-videos. This is accomplished by enforcing the external sound clips and the internal acoustic modality to share the same acoustic concept-level space. Notably, this scheme is applicable to enhance other modality representation, like visual and textual ones.

In order to intelligently route micro-videos to the target users, we present a multimodal sequential learning scheme. To capture the users' dynamic and diverse interest, we encode their historical complicated interaction sequences into a temporal graph and then design a novel temporal graph-based long short-term memory (LSTM) network to model it. Afterwards, we estimate the click probability via calculating the similarity between the users' interest representation and the embedding of the given micro-video. Considering that users' interest is multi-level, we introduce a user matrix to enhance the user interest modeling by incorporating their "like" and "follow" information. And at this step, we also get a click probability with respect to users' more precise interest information. Analogously, since we know the sequence of users' disliked micro-videos, another temporal graph-based LSTM is built to characterize users' uninterested information, and the other click probability can be estimated based on true negative samples. We can thus obtain a click probability regarding users' uninterested information. Finally, the weighted sum of the above three probability scores is set as our final prediction result.

1.5   BOOK STRUCTURE

In this book, we present an in-depth introduction to multimodal learning toward micro-video understanding, and a comprehensive literature survey of all the important research topics and latest state-of-the-art methods in the area. It is suitable for students, researchers, and practitioners who are interested in multimodal learning. It is worth emphasizing that the multimodal learning methods presented in this book are applicable to other fields owing multi-aspect data, like web image analysis, visual question answering, and user profiling across multiple social networks.

The remainder of this book consists of six chapters. Chapter 2 introduces the three micro-video benchmark datasets for three practical tasks. Chapter 3 describes a multimodal transductive learning framework to tackle the information sparsity and low-quality problem. We theoretically derive its closed-form solution and practically apply it to the micro-video popularity prediction. In Chapter 4, we present a series of multi-modal cooperative learning methods toward characterizing the explicit correlations among different modalities, such as consistent and complementary relationship. These methods are verified over the task of venue category estimation of micro-videos. In Chapter 5, we devise a deep transfer learning model by harnessing the external sound knowledge to compensate the acoustic modality in micro-videos. This is a robust method to address the low-quality problem. Following that, in Chapter 6, we study the multimodal sequential property of micro-videos and testify its effectiveness over the task of micro-video routing. We finally conclude this book and figure out the future research directions in Chapter 7.

CHAPTER 2

Data Collection

In this book, we have three micro-video datasets corresponding to the tasks of popularity prediction, venue category estimation, and micro-video routing, respectively. In this chapter, we detail them one by one.

2.1 DATASET I FOR POPULARITY PREDICTION

The first micro-video data collection, dubbed Dataset I, was crawled from one of the most prominent micro-video sharing social networks, Vine. The reason we chose Vine is because in addition to the historical uploaded micro-videos, it also archives users' profiles and their social connections.

In particular, we first randomly selected 10 active Vine users from Rankzoo,[1] which provides the top 1,000 active users of Vine, as the seed users. Considering that these seed users may have millions of followers, we practically only retained the first 1,000 returned followers for each seed user to improve the crawling efficiency. We then adopted the breadth-first strategy to expand our user set by gathering their followers. This is accomplished with the help of the public Vine API.[2] We terminated our expansion after three layers. After three layers of crawling, we harvested a densely connected user set consisting of 98,166 users as well as 120,324 following relationships among users. For each user, his/her brief profile was crawled, containing full name, description, location, follower count, followee count, like count, post count, and loop count of all posted videos. Besides, we also collected the timeline (the micro-video posting history, including the repostings from others) of each user between July 1 and October 1, 2015. Finally, we obtained 1.6 million video postings, including a total number of 303,242 unique micro-videos with a total duration of 499.8 h. In Figure 2.1, we show the procedure of the Dataset I collection.

To measure the popularity of micro-videos, we considered four popularity-related indicators as shown in Figure 2.2, namely, the number of comments (n_comments), the number of likes (n_likes), the number of reposts (n_reposts), and the number of loops/views (n_loops) to measure the popularity of micro-videos. Figure 2.3 illustrates the proportion of micro-videos regarding each of the four indicators in our dataset; it is noted that each distribution is different, and each measures one aspect of popularity. In order to comprehensively and precisely measure the popularity of each micro-video, y_i, we linearly fuse all four indicators as the popularity

---

[1]https://rankzoo.com/vine_users
[2]https://github.com/davoclavo/vinepy

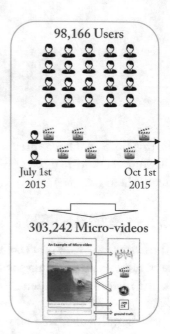

Figure 2.1: Crawling strategies of Dataset I from the Vine service.

ground truth:

$$y_i = \frac{(n\_reposts + n\_comments + n\_likes + n\_loops)}{4}. \tag{2.1}$$

2.2 DATASET II FOR VENUE CATEGORY ESTIMATION

As mentioned in Section 2.1, we obtained 98,166 users through the breadth-first crawling strategy. For each user, we crawled all his/her historical data, without time constraints, including published videos, video descriptions, and venue information if available. In such a way, we harvested 2 million micro-videos, however, only about 24,000 micro-videos contain Foursquare check-in information. After removing the duplicate venue IDs, we further expanded our video set by crawling all videos in each venue ID with the help of Vine API. This eventually yielded a dataset of 276,264 videos distributed in 442 Foursquare venue categories. Each venue ID was mapped to a venue category via the Foursquare API, which serves as the ground truth. The crawling strategy of Dataset II is visualized in Figure 2.4. Foursquare organizes its venue categories into a four-layer hierarchical structure, as shown in Figure 1.2, with 341, 312, and 52 leaf nodes in the second layer, third layer, and fourth layer, respectively. The top-layer of this structure contains ten non-leaf nodes (coarse venue categories). To visualize the coverage and representativeness of our collected micro-videos, we plotted and compared the distribution curves over the

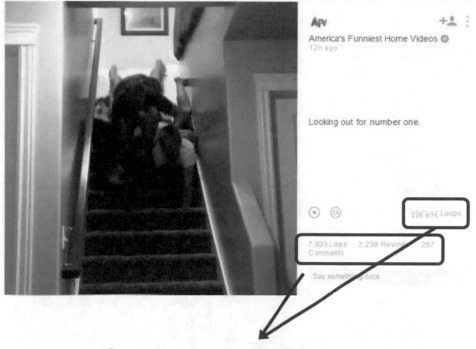

Figure 2.2: Illustration of four indicators used in this book.

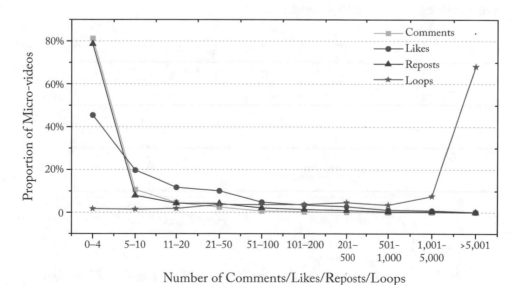

Figure 2.3: Distribution of the number of comments, likes, reposts, and loops of micro-videos in Dataset I collected from the Vine website.

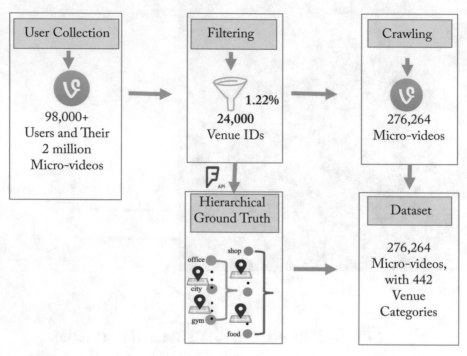

Figure 2.4: Distribution of the number of comments, likes, reposts, and loops of micro-videos in Dataset I collected from the Vine website.

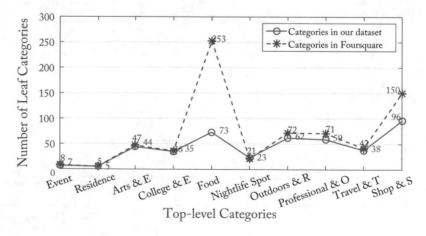

Figure 2.5: Top-level venue category distribution in terms of the number of leaf nodes.

number of leaf categories between our dataset and the original structure, as shown in Figure 2.5. It is worth mentioning that the number of leaf categories distributed in Foursquare is extremely unbalanced. For instance, the "Food" category has 253 leaf nodes, while the "Residence" only contains 5 leaf nodes. Accordingly, the distribution of our crawled videos over the top-layer categories also shows such unbalance, as displayed in Table 2.1.

On the other hand, we observed that some leaf categories contain only a small number of micro-videos. For instance, "Bank/Financial" only consists of three samples in our dataset, which is hard to train a robust classifier. We hence removed the leaf categories with less than 50 micro-videos. At last, we obtained 270,145 micro-videos distributed in 188 Foursquare leaf categories. Table 2.2 lists the top 5 leaf categories with the most and the least micro-videos, respectively.

We observed that the acoustic and textual modalities are missing in some micro-videos. More precisely, there are 169 and 24,707 micro-videos with missing acoustic and textual modality, respectively. Information missing is harmful for most machine learning performance [1, 12], including the models for venue category estimation. To alleviate such a problem, we cast the data completion task as a matrix factorization problem [48]. In particular, we first concatenated the features from three modalities in order, which naturally constructed an original matrix. We then

Table 2.1: Number of micro-videos in each of the 10 categories in the first layer

Top-Layer Category	Number	Top-Layer Category	Number
Outdoors and Recreation	93,196	Shop and Service	10,976
Arts and Entertainment	88,393	Residence	8,867
Travel and Transport	24,916	Nightlife Spot	8,021
Professional and Other	18,700	Food	6,484
College and Education	12,595	Event	1,047

Table 2.2: Leaf categories with the most and the least of micro-videos

Leaf Category with the Most Videos	Number	Leaf Category with the Least Videos	Number
City	30,803	Bakery	53
Theme Park	16,383	Volcano	51
Neighborhood	15,002	Medical	51
Other Outdoors	10,035	Classroom	51
Park	10,035	Toy and Games	50

applied the matrix factorization technique [37] to factorize this original matrix into two latent matrices with 100 latent features, such that the empirical errors between the production of these 2 latent matrices and the original matrix are as small as possible. The entries in the 2 latent matrices are inferred by the observed values in the original matrix only, and over-fitting is avoided through a regularized model.

2.3   DATASET III FOR MICRO-VIDEO ROUTING

We evaluate our proposed multimodal sequential learning methods for the task of micro-video routing on two public micro-video, Dataset III-1 and Dataset III-2.

Dataset III-1. This dataset is released by the Kuaishou Competition in ChinaMM2018 conference,[3] which aims to infer users' click probabilities for new micro-videos. In this dataset, there are multiple interactions between users and micro-videos, such as click, not click, like, and follow. Particularly, not click means the user did not click the micro-video after previewing its thumbnail. Moreover, each behavior is associated with a timestamp, which records when the behavior happens. We have to mention that the timestamp has been processed such that the absolute time is unknown, but the sequential order can be obtained according to the timestamp. For each micro-video, the contest organizers have released its 2,048-d visual embedding of its thumbnail. Among the large-scale dataset, we randomly selected 10,000 users and their 3,239,534 interacted micro-videos to construct the Dataset III-1.

Dataset III-2. This dataset is constructed by [4] for micro-video click-through prediction. It consists of 10,986 users, 1,704,880 micro-videos, and 12,737,619 interactions. Different from Dataset III-1, Dataset III-2 only contains the "click" and "not click" behaviors. Each micro-video in Dataset III-2 is represented by a 512-d visual embedding vector extracted from its thumbnail and associated with a category label, and each user's behavior is linked with a processed timestamp.

The statistics of the above two datasets are summarized in Table 2.3. The reported experimental results of micro-video routing in this book are based on these two datasets. Specifically, we set the first 80% of a user's historical accessed micro-videos as the training set and the rest of 20% as the testing one in the Dataset III-1. As for Dataset III-2, we utilized the same setting with [28]. It is worth mentioning that we adopted the Principal Component Analysis (PCA) [169] to reduce the visual embedding vector of a micro-video to 64 dimension.

2.4   SUMMARY

To justify our proposed models and its three practical application scenarios, in this chapter, we introduce three micro-video datasets. In particular, we construct Dataset I and II for popularity prediction and venue category estimation, respectively. In addition, we leverage the publicly released micro-video datasets, namely Dataset III-1 and III-2, to testify the task of micro-video

[3] http://mm.ccf.org.cn/chinamm/2018/

Table 2.3: Statistics of the two datasets

Dataset	Dataset I	Dataset II
# Users	10,000	10,986
# Items	3,239,534	1,704,880
# Interactions	13,661,383	12,737,619
# Interaction types	4	2
# Average interactions per user	1,366.14	1,159.44
# Average interactions per item	4.28	7.47
# Average clicked items per user	277	218
# Interactions in training set	10,931,092	8,970,310
# Interactions in test set	2,730,291	3,767,309

routing. We have released all the involved codes, parameter settings, and datasets involved in this book to facilitate other researchers in the community of micro-video understanding.[4]

---

[4]https://ilearn2019.wixsite.com/microvideo.

CHAPTER 3

Multimodal Transductive Learning for Micro-Video Popularity Prediction

3.1 BACKGROUND

Despite their shortness, micro-videos generally outline a relatively simple but complete story to audiences. Within a limited time interval, producers also attempt to condense and maximize what they want to say, thereby to create more attractive stories. Compared with traditional long videos like the ones in Youtube, micro-videos are produced to satisfy a fast-paced modern society, which makes micro-videos appear to be more social-oriented. Therefore, micro-videos are much easier to be spread. Popular micro-videos have enormous commercial potential in many ways, such as online marketing and brand tracking. In fact, the popularity prediction of traditional UGCs including tweets, web images, and long videos, has achieved good theoretical underpinnings and great practical success. However, little research has thus far been conducted to predict the popularity of the bite-sized videos. In this chapter, we work toward solving the problem of popularity prediction of micro-videos posted on social networks.

3.2 RESEARCH PROBLEMS

Since micro-videos are produced with the aim of rapid spreading and sharing among users, these videos bring more intrinsic relations with social networks that differ from traditional long videos. Therefore, it makes predicting the popularity of micro-videos a non-trivial task due to the following facts.

(1) **Heterogeneous**. Due to the short duration of micro-videos, each modality can only provide limited information, the so-called modality limitation. Fortunately, micro-videos always involve multiple modalities, namely, social, visual, acoustic, and textual[1] modalities. In a sense, these modalities are co-related rather than independent and essentially characterize the same micro-videos. Therefore, the major challenge lies on how to effectively fuse micro-videos' heterogeneous clues from multiple modalities [162, 163, 189]. The most naive strategies are early

---

[1]Micro-videos are usually associated with certain textual data, such as video descriptions given by the video owners.

fusion and late fusion [149]. They, however, fail to account for the relatedness among multiple modalities. Therefore, it is important to take modality relatedness into consideration.

(2) **Interconnected**. Heterogeneous features extracted from different modalities show different aspects of micro-videos, which are complementary to each other. In this case, it will be beneficial to develop an effective approach to finding the interconnected patterns shared by all views. However, due to the restrictions brought by micro-video producers and platforms, the additional information associated with a micro-video, textual description, for example, suffers from the diverse or unstructured nature, causing the features extracted from certain views unavailable in many situations. For example, according to our statistics over around 2 million Vine micro-videos, as reported in [24], more than 11% of micro-videos do not provide textual descriptions. In contrast, micro-video content itself ensures a steady information source to enable popularity prediction. Thus, to compensate for this limitation, micro-video content features are considered to be an indispensable component for a more descriptive and predictive analysis on the one hand, and it is necessary to exploit the complementarity between different views to learn the latent interconnected patterns to address the incomplete problem on the other hand.

(3) **Noisy**. Originating from certain external factors in reality, various types of noises make the real underlying data structure hidden in the observed data. For example, micro-videos are often captured by users with hand-held mobile devices which easily result in poor video quality, such as low-resolution, wobbly frames, constrained lighting conditions, and background noise. Besides, textual descriptions related to micro-videos may be noisy and uncorrelated. The aforementioned challenges drive us to build a robust model to explore the intrinsic structure property embedded in data by inferring meaningful features and alleviating the impact of noisy ones.

3.3   FEATURE EXTRACTION

It is apparent that both the publisher influence and content influence contribute to the popularity of UGCs. In particular, we characterized the publisher influence via the social modality, and the content influence via visual, acoustic and textual modalities. For content influence, we first examined the popular micro-videos in Dataset I and proposed three common characteristics of online micro-videos. For each characteristic, we then explained the insights, and transformed it into a set of features for video representation. Finally, we developed a rich set of popularity-oriented features from each modality of micro-videos in Dataset I.

3.3.1   OBSERVATIONS

Universal Appeal. The subjects of widely popular micro-videos cannot be something that can only be appreciated by a small group of people. Therefore, the topics and objects contained in micro-videos should be something common so that to be interpreted the same way across people and cultures. To capture this characteristic, we extracted Sentence2Vector feature from the textual modality and deep object feature from the visual one.

Emotional Content. People are naturally drawn to things that arouse their emotions. Micro-videos showing funny animals or lovely babies make people feel urge to share them to express the same emotions. As a result, micro-videos that are highly emotional are more likely to be shared. Therefore, we extracted textual sentiment, visual sentiment features for each video as well as several acoustic features, which is widely used in emotion recognition in music [171].

High Quality and Aesthetic Design. When people share information on social networks, people are actually showing a little piece of themselves to their audience. Therefore, high quality and aesthetic design of the content, which could reflect the taste of people, is another important characteristic of popular micro-videos. Color histogram, aesthetic feature, and visual quality feature were thus extracted to encode such characteristic. In addition, the acoustic features we extracted are frequently used in music modeling, which could help to detect music in the audio track of micro-videos [88].

3.3.2 SOCIAL MODALITY

It is intuitive that micro-videos posted by users, who has more followers or has a verified account, are more likely to be propagated, and thus tend to receive a higher number of audiences. To characterize the influence of micro-video publishers, we developed the following publisher-centric features for micro-videos.

- **Follower/Followee Count**. The number of followers and followees of the given micro-video publisher.

- **Loop Count**. The total number of loops received by all the posts of the publisher.

- **Post Count**. The number of posts generated by the publisher.

- **Twitter Verification**. A binary value indicating whether the publisher has been verified by Twitter.[2]

3.3.3 VISUAL MODALITY

Due to the short length of micro-videos, the visual content is usually highly related to a single theme, which enables us to only employ a few key frames to represent the whole micro-video. Inspired by this, we extracted the visual features from certain key frames. The mean pooling was performed across all the key frames to create a fixed-length vector representation of each micro-video.

Color Histogram It has been found that most basic visual features (i.e., intensity and the mean value of different color channels in HSV space) except color histogram, have little correlation with popularity [77]. Color histogram has outstanding correlation due to the fact that striking

---

[2]A Vine account can be verified by Twitter, if it is linked to a verified Twitter account.

colors tend to catch users' eyes. Therefore, we only extracted color histogram as the basic visual feature to characterize popular micro-videos. To reduce the size of color space, we grouped the color space into 50 distinct colors, which results in a 50-d vector for each frame.

Object Features   It has been studied that popular UGCs are strongly correlated with the objects contained in the videos [54]. We believe that the presence of certain objects affect micro-videos' popularity. For example, micro-videos with "cute dogs" or "beautiful girls" are more likely to be popular than those with "desks" and "stones." We thus employed the deep convolutional neural networks (CNNs) [82], a powerful model for image recognition problems [188], to detect objects in micro-videos. Specifically, we applied the well-trained AlexNet deep neural network (DNN) provided by the Caffe software package [71] to the input key frames. The output of the fc7 layer and the final 1,000-way softmax layer in AlexNet is a probability distribution over the 1,000 class labels predefined in ImageNet. We treat them as our feature representation of each frame. In the end, a mean pooling was performed over the frames to generate a single 4,096-d vector and 1,000-d vector for each micro-video.

SentiBank Features   We performed the sentiment analysis of the visual modality due to that the sentiment of UGCs has been proven to be strongly correlated with their popularity [54]. In particular, we extracted the visual sentiment features based on the deep CNNs model which was trained on the SentiBank dataset [11]. SentiBank contains 2,089 concepts and each of them invokes specific sentiments such as "cute girls" and "funny animals." Therefore, after mean pooling among keyframes, each micro-video is represented by a 2,089-d vector.

Aesthetic Features   Aesthetic features are a set of handful selected features related to the principles of the nature and appreciation of beauty, which have been studied and found to be effective in popularity prediction [36]. Intuitively, micro-videos that are objectively aesthetic are more likely to be popular. We employed the released tool[3] [10] to extract the following aesthetic features: (a) dark channel feature; (b) luminosity feature; (c) sharpness; (d) symmetry; (e) low depth of field; (f) white balance; (g) colorfulness; (h) color harmony, and (i) eye sensitivity, at 3×3 grids over each key frame. We then calculated: (a) normalized area of dominant object and (b) normalized distances of centroid of dominant objects with respect to four stress points at frame level. In the end, we obtained 149-d aesthetic features for each micro-video.

Visual Quality Assessment Features   It is important that the visual quality of popular contents are maintained at an acceptable level, given rising consumer expectations of the quality of multimedia content delivered to them [140]. In particular, we employed the released tool[4] to extract the micro-videos quality features based on the motion and spatio-temporal information, which have been proven to correlate highly with human visual judgments of quality. This results in a 46-d features.

---

[3]http://www.ee.columbia.edu/~subh/Software.php
[4]http://live.ece.utexas.edu/

3.3.4   ACOUSTIC MODALITY

Acoustic modality usually works as an important complement to visual modality in many video-related tasks, such as video classification [175]. In fact, audio channels embedded in the micro-videos may also contribute to the popularity of micro-videos to a large extent. For example, the audio channel may indicate the quality of a given micro-video and convey rich background information about the emotion as well as the scene contained in the micro-video, which significantly affects the popularity of a micro-video. The acoustic information is especially useful for the cases where the visual features could not carry enough information. Therefore, we adopted the following widely used acoustic features, i.e., mel-frequency cepstral coefficients (MFCC) [88] and Audio-Six (i.e., Energy Entropy, Signal Energy, Zero Crossing Rate, Spectral Rolloff, Spectral Centroid, and Spectral Flux [171]). These features are frequently used in different audio-related tasks, such as emotion detection and music recognition. We finally obtained a 36-d acoustic feature vector for each micro-video.

3.3.5   TEXTUAL MODALITY

Micro-videos are usually associated with textual modality in the form of descriptions, such as "when Leo finally gets the Oscar" and "Puppy dog dreams," which may precisely summarize the micro-videos. Such summarization may depict the topics and sentiment information regarding the micro-videos, which has been proven to be of significance in online article popularity prediction [9].

Sentence2Vector   We found that the popular micro-videos are sometimes related to the topics of the textual descriptions. This observation propels us to conduct content analysis over the textual descriptions of micro-videos. Considering the short-length of descriptions, to perform content analysis, we employed the state-of-the-art textual feature extraction tool Sentence2Vector,[5] which was developed on the basis of work embedding algorithm Word2Vector [115]. In this way, we extracted 100-d features for video descriptions.

Textual Sentiment   We also analyze the sentiments over text, which has been proven to play an important role in popularity prediction [8]. With the help of the Sentiment Analysis tool in Stanford CoreNLP tools,[6] we assigned each micro-video a sentiment score ranging from 0–4 and they correspond to *very negative*, *negative*, *neutral*, *positive*, and *very positive*, respectively.

---

[5]https://github.com/klb3713/sentence2vec
[6]http://stanfordnlp.github.io/CoreNLP/

3.4   RELATED WORK

3.4.1   POPULARITY PREDICTION

Significant efforts were devoted to exploring the popularity prediction of items such as text [5, 107], images [16, 54, 112], and videos [18, 85, 100, 156] due to their potential value in business [158, 170].

For the task of predicting the popularity of text, most methods tend to explore the textual content itself and the correlation between popularity and the social context. For example, Ma et al. [107] proposed to predict the popularity of new hashtags on Twitter by extracting 7 content features from both hashtags and tweets and 11 contextual features from the social graphs formed by users.

As to the image popularity, content-based image features, context features, and social context features are generally exploited to predict image popularity. For example, Khosla et al. [77] explored the relative significance of individual features involving multiple visual features, such as color, gradient, texture, and the presence of objects, as well as various social context features, such as the number of normalized views or contacts. Totti et al. [155] presented a complementary analysis on how the aesthetic properties, such as brightness, contrast and sharpness, and semantics contribute to image popularity. Gelli et al. [54] proposed to combine user features and three context features together with image sentiment features to better predict the popularity of social images.

When it comes to video popularity prediction, analogous to images, videos also integrate different information channels, like visual, acoustic, social, and textual modalities. The majority of studies focus on investigating the factors that determine the popularity of videos [18, 85, 156, 173]. For example, Cha et al. [18] conducted a large-scale data-driven analysis to uncover the latent correlations between video popularity and UGC. Li et al. [85] proposed using both video attractiveness and social context as inputs to predict video views on online social networks. Trzcinski et al. [156] employed temporal and visual cues to predict the popularity of online videos. The tasks above share the same thing—they do not describe each item based on its content only; instead, they mine multiple views of context information related to the item and social cues from the users to improve the prediction performance. In addition, some researchers [24, 38, 85] have explored the improvement of video popularity prediction by fusing information introduced by different patterns. To overcome the ineffectiveness of traditional models, such as autoregressive integrated moving average (ARIMA), multiple linear regression (MLR), and k-nearest neighbors (kNN), when predicting the popularity of online videos, Li et al. [85] introduced a novel propagation-based popularity prediction method by considering both video intrinsic attractiveness and the underlying propagation structure. Roy et al. [138] used transfer learning to model the social prominence of videos, in which an intermediate topic space is constructed to connect the social and video domains. Ding et al. [38] developed a dual sentimental Hawkes process (DSHP) for video popularity prediction, which not only takes sentiments in video popularity propagation into account but also reveals more underlying factors that determine the popularity of a video.

However, the aforementioned studies do not consider the combined impact of heterogeneous, interconnected, and noisy data. In contrast, our proposed scheme not only pursues a solid fusion of heterogeneous multi-view features based on the complementary characteristics but also concentrates on exploiting the advantages of the low-rank representation to learn robust features within the incomplete and noisy data. As a complement, we aim to timely predict the popularity of a given micro-video even before it get published by proposing a novel multi-modal learning scheme.

3.4.2 MULTI-VIEW LEARNING

Technically speaking, traditional multimodal fusion approaches consist of early fusion and late fusion. Early fusion approaches, such as [42, 146], typically concatenate the unimodal features extracted from each individual modality into a single representation to adapt to the learning setting. Following that, one can devise a classifier, such as a neural network, treating the overall representation as the input. However, these approaches generally overlook the obvious fact that each view has its own specific statistical property and ignore the structural relatedness among views. Hence, it fails to explore the modal correlations to strengthen the expressiveness of each modality and further improve the capacity of the fusion method. Late fusion performs the learning directly over unimodal features, and then the prediction scores are fused to predict the venue category, such as averaging [144], voting [118] and weighting [134]. Although this fusion method is flexible and easy to work, it overlooks the correlation in the mixed feature space.

In contrast to the early and late fusion, as a new paradigm, multi-view learning exploits the correlations between the representations of the information from multiple modalities to improve the learning performance. It can be classified into three categories: co-training, multiple kernel learning, and subspace learning.

Co-training [31] is a semi-supervised learning technique which first learns a separate classifier for each view using the labeled examples. It maximizes the mutual agreement on two distinct views of the unlabeled data by alternative training. Many variants have since been developed. Instead of committing labels for the unlabeled examples, Nigam et al. [125] proposed a co-EM approach to running EM in each view and assigned probabilistic labels to the unlabeled examples. To resolve the regression problems, Zhou and Li [208] employed two k-nearest neighbor regressors to label the unknown instances during the learning process. More recently, Yu et al. [186] proposed a Bayesian undirected graphical model for co-training through the Gaussian process. The success of the co-training algorithms relies on three assumptions: (a) each view is sufficient to estimate on its own; (b) it is probable that a function predicts the same labels for each view feature; and (c) the views are conditionally independent of the given label. However, these assumptions are too strong to satisfy in practice, especially for the micro-videos with different modalities, whereby the information in each modality is insufficient to generate the same label prediction.

Multiple Kernel Learning [45] leverages a predefined set of kernels corresponding to different views and learns an optimal linear or nonlinear combination of kernels to boost the performance. Lanckriet et al. [83] constructed a convex Quadratically Constrained Quadratic Program by conically combining the multiple kernels from a library of candidate kernels and applied the method to several applications. To extend this method to a large-scale dataset, Bach et al. [4] took the dual formulation as a second-order cone programming problem and developed a sequential minimal optimization algorithm to obtain the optimal solution. Further, Ying and Campbell [184] used the metric entropy integrals and pseudo-dimension of a set of candidate kernels to estimate the empirical Rademacher chaos complexity.

Subspace learning [161] obtains a latent subspace shared by multiple views by assuming that the input views are generated from this subspace. The dimensionality of the subspace is lower than that of any input view, so the subspace learning alleviates the "curse of dimensionality." The canonical correlation analysis (CCA) [65] is straightforwardly applied to select the shared latent subspace through maximizing the correlation between the views. Since the subspace is linear, it is impossible to apply CCA to the real-world datasets exhibiting nonlinearities. To compensate for this problem, Akaho [1] proposed a kernel variant of CCA, namely KCCA. Diethe et al. [37] proposed the Fisher Discriminant Analysis using the label information to find the informative projections, more informative in the supervised learning settings. Recently, Zhai et al. [187] studied the multi-view metric learning by constructing embedding projections from multi-view data to a shared subspace. Although the subspace learning approaches alleviate the "curse of dimensionality," the dimensionality of subspace changes along with the task.

Overall, compelling success has been achieved by multi-view learning models on various problems, such as categorization [145, 150], clustering [19, 51] and multimedia retrieval [91, 92]. However, to the best of our knowledge, limited efforts have been dedicated to applying multi-view learning in the context of micro-video understanding, which is the major concern of our work.

3.4.3   LOW-RANK SUBSPACE LEARNING

In recent years, low-rank representation [95, 198–200] has been considered as a promising technique for exploring the latent low-dimensional representation embedded in the original space. Low-rank subspace learning has been applied to a wide range of machine learning tasks, including matrix recovery [210], image classification [106, 197], subspace segmentation [96], and missing modality recognition [40].

Robust principal component analysis (RPCA) [74] is a popular low-rank matrix recovery method for high-dimensional data processing. This method aims to decompose a data matrix into a low-rank matrix and a sparse error matrix. To promote the discriminative ability of the original RPCA and improve the robust representation of corrupted data, Chen et al. [22]

presented a novel low-rank matrix approximation method with a structural incoherence constraint, which decomposes the raw data into a set of representative bases with associated sparse error matrices. Based on the principle of self-representation, Liu et al. [95] proposed the low-rank representation (LRR) method to search for the lowest-rank representation among all the candidates. To overcome the incompetence of LRR in handling unobserved, insufficient, and extremely noisy data, Liu and Yan [96] further developed an advanced version of LRR, called latent low-rank representation (LatLRR), for subspace segmentation. Zhang et al. [196] proposed a structured low-rank representation method for image classification, which constructs a semantic-structured and constructive dictionary by incorporating class label information into the training stage. Zhou et al. [205] provided a novel supervised and low-rank-based discriminative feature learning method that integrates LatLRR with ridge regression to minimize the classification error directly.

To handle data that are generated from multiple views in many real-world applications, some multi-view low-rank subspace learning methods have been developed to search for a latent low-dimensional common subspace such that it can capture the commonality among all the views. For example, Xia et al. [176] proposed to construct a transition probability matrix from each view and then recover a shared low-rank transition probability matrix via low-rank and sparse decomposition. Liu et al. [101] presented a novel low-rank multi-view matrix completion (lrMMC) method for multi-label image classification, where a set of basic matrices are learned by minimizing the reconstruction errors and the rank of the latent common representation. In the case that the view information of the testing data is unknown, Ding and Fu [39] proposed a novel low-rank common subspace (LRCS) algorithm in a weakly supervised setting, where only the view information is employed in the training phase. In [41], a dual low-rank decomposition model was developed to learn a low-dimensional view-invariant subspace. To guide the decomposition process, two supervised graph regularizers were considered to separate the class structure and view structure. Li et al. [86] proposed a novel approach, named low-rank discriminant embedding (LRDE), by considering the correlations between views and the geometric structures contained within each view simultaneously. These multi-view low-rank learning approaches have been proven to be effective when different feature views are complementary to each other.

Although low-rank representation enables an effective learning mechanism in exploring the low-rank structure in noisy datasets [177], only a limited amount of low-rank models have been developed to address the popularity prediction in social networks. The prediction of video popularity can be considered as a standard regression problem. To the best of our knowledge, one of the most related work to our approach is introduced in [202], in which a multi-view low-rank regression model is presented by imposing low-rank constraints on the multi-view regression model. However, in that work, the structure and relations among different views were ignored. To overcome this drawback, we propose to learn a set of view-specific projections by maximizing the total correlations among views to map multi-view features into a common space. Another difference is that the lowest-rank representation is adaptively obtained by a low-rank constraint, which is approximated by the trace norm rather than being specified in advance.

3.5   NOTATIONS AND PRELIMINARIES

We first declare several notations. In particular, we employ bold capital letters (e.g., $\mathbf{X}$) and bold lowercase letters (e.g., $\mathbf{x}$) to denote matrices and vectors, respectively. We use non-bold letters (e.g., x) to represent scalars, and Greek letters (e.g., β) as parameters. If not clarified, all vectors are in column form. Moreover, given a matrix $\mathbf{M} \in R^{N \times D}$, its i-th row and the i-th column of matrix $\mathbf{M}$ are denoted by $\mathbf{m}^i$ and $\mathbf{m}_i$, respectively. The $\mathcal{L}_{p,q}$-norm of matrix $\mathbf{M}$ is defined as

$$\|\mathbf{M}\|_{p,q} = \left[\sum_{i=1}^{N} \left(\sum_{j=1}^{D} |M_{ij}|^p \right)^{q/p} \right]^{1/q}, \tag{3.1}$$

where M_{ij} is the (i, j)-th element of matrix $\mathbf{M}$. By assigning different values to p and q, there are several regularization terms, which are stated as follows.

The $\mathcal{L}_1$-norm is defined when $p = q = 1$,

$$\|\mathbf{M}\|_1 = \sum_{i=1}^{N} \|\mathbf{m}^i\|_1 . \tag{3.2}$$

The Frobenius norm $\mathcal{L}_F$ is defined when $p = q = 2$,

$$\|\mathbf{M}\|_F = \sqrt{\sum_{i=1}^{N} \sum_{j=1}^{D} \|M_{ij}\|^2}. \tag{3.3}$$

The trace norm of matrix $\mathbf{M}$ is defined as

$$\|\mathbf{M}\|_* = \sum_i \delta_i(\mathbf{M}), \tag{3.4}$$

where $\sum_i \delta_i(\mathbf{M})$ is the sum of singular values of matrix $\mathbf{M}$.

Our proposed model targets at reasoning from observed training micro-videos to testing ones. Such prediction belongs to transductive learning, in which both labeled samples as well as unlabeled samples are available for training. It hence obtains better performance. In contrast, inductive model is reasoning from observed training cases to general rules, which are then applied to the test cases.

3.6   MULTIMODAL TRANSDUCTIVE LEARNING

Without loss of generality, suppose we have N labeled samples and M unlabeled samples with $K \geqslant 2$ modalities. It is worth noting that the unlabeled samples also serve as testing samples. Z_k stands for the number of features extracted from the k-th modality. Then the k-th modality can be represented as $\mathbf{X}_k \in \mathbb{R}^{(N+M)\times Z_k}$. The popularity of all the videos are denoted by $\mathbf{y} = \{y_1, y_2, \ldots, y_N\}^T \in \mathbb{R}^N$. Let $\mathbf{f} = \{f_1, f_2, \ldots, f_N, f_{N+1}, f_{N+2}, \ldots, f_{N+M}\}^T \in \mathbb{R}^{N+M}$ stand for the predicted results regarding popularity for all samples, including the labeled and unlabeled ones. We aim to jointly learn the common space $\mathbf{X}_0 \in \mathbb{R}^{(N+M)\times Z_0}$ shared by multiple modalities and the popularity for the M unlabeled micro-videos.

We present a novel **T**ransductive **M**ulti-mod**AL** **L**earning approach, TMALL for short, to predicting the popularity of micro-videos. As illustrated in Figure 3.1, we first crawl a representative micro-video dataset from Vine and develop a rich set of popularity-oriented features from multi-modalities. We then perform multi-modal learning to predict the popularity of micro-videos, which seamlessly takes the modality relatedness and modality limitation into account by utilizing a common space shared by all modalities. We assume that there exists an optimal common space, which maintains the original intrinsic characteristics of micro-videos in the original spaces. In light of this, all modalities are forced to be correlated. Meanwhile, micro-videos with different popularity can be better separated in such optimal common space, as compared to that of each single modality. In a sense, we alleviate the modality limitation problem. It is worth mentioning that, in this work, we aim to predict how popular a given micro-video will be when the propagation is stable rather than when the given micro-video would be popular.

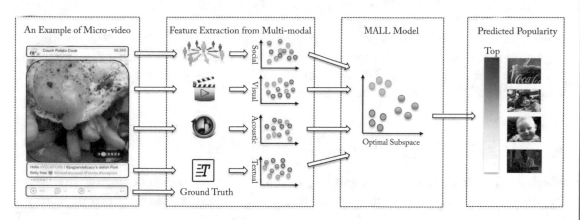

Figure 3.1: Micro-video popularity prediction via our proposed TMALL model.

3.6.1   OBJECTIVE FORMULATION

It is apparent that different modalities may contribute distinctive and complementary information about micro-videos. For example, textual modality gives us hints about the topics of the given micro-video; acoustic and visual modalities may, respectively, convey location and situation of micro-videos, and user modality demonstrates the influence of the micro-video publisher. These clues jointly contribute to the popularity of a micro-video. Obviously, due to the noise and information insufficiency of each modality, it may be suboptimal to conduct learning directly from each single modality separately. In contrast, we assume that there exists an optimal latent space, in which micro-videos can be better described. Moreover, the optimal latent space should maintain the original intrinsic characteristics conveyed by multi-modalities of the given micro-videos. Therefore, we penalize the disagreement of the normalized Laplacian matrix between the latent space and each modality. In particular, we formalize this assumption as follows: Let $\mathbf{S}_k \in \mathbb{R}^{(N+M) \times (N+M)}$ be the similarity matrix,[7] which is computed by the Gaussian similarity function as follows:

$$
S_k(i, j) = \begin{cases} \exp\left(-\dfrac{\left\|\mathbf{x}_k^i - \mathbf{x}_k^j\right\|^2}{2\sigma_k^2}\right) & , \text{ if } i \neq j, \\ 0 & , \text{ if } i = j, \end{cases} \tag{3.5}
$$

where $\mathbf{x}_k^i$ and $\mathbf{x}_k^j$ are the micro-video pairs in the k-th modality space. Thereinto, the radius parameter σ_k is simply set as the median of the Euclidean distances over all video pairs in the k-th modality. We then derive the corresponding normalized Laplacian matrix as follows:

$$
\mathbf{L}(\mathbf{S_k}) = \mathbf{I} - \mathbf{D}_k^{-\frac{1}{2}} \mathbf{S}_k \mathbf{D}_k^{-\frac{1}{2}}, \tag{3.6}
$$

where $\mathbf{I}$ is a $(N + M) \times (N + M)$ identity matrix and $\mathbf{D}_k \in \mathbb{R}^{(N+M) \times (N+M)}$ is the diagonal degree matrix, whose (u, u)-th entry is the sum of the u-th row of $\mathbf{S}_k$. Since $S_k(i, j) > 0$, we can derive that $tr(\mathbf{L}(\mathbf{S_k})) > 0$. We thus can formulate the disagreement penalty between the latent space and the original modalities as

$$
\sum_{k=1}^{K} \left\| \frac{1}{tr(\mathbf{L}(\mathbf{S_0}))} \mathbf{L}(\mathbf{S_0}) - \frac{1}{tr(\mathbf{L}(\mathbf{S_k}))} \mathbf{L}(\mathbf{S_k}) \right\|_F^2, \tag{3.7}
$$

where $tr(\mathbf{A})$ is the trace of matrix $\mathbf{A}$ and $\|\cdot\|_F$ denotes the Frobenius norm of matrix. In addition, inspired by [164], considering that similar micro-videos attempt to have similar popularity in the latent common space, we adopt the following regularizer:

$$
\frac{1}{2} \sum_{m=1}^{N+M} \sum_{n=1}^{N+M} \left(\frac{f(\mathbf{x}_0^m)}{\sqrt{D_0(\mathbf{x}_0^m)}} - \frac{f(\mathbf{x}_0^n)}{\sqrt{D_0(\mathbf{x}_0^n)}} \right)^2 S_0(m, n) = \mathbf{f}^T \mathbf{L}(\mathbf{S_0}) \mathbf{f}. \tag{3.8}
$$

Based upon these formulations, we can define the loss function that measures the empirical error on the training samples. As reported in [123], the squared loss usually yields good performance as other complex ones. We thus adopt the squared loss in our algorithm for simplicity and efficiency. In particular, since we do not have the labels for testing samples, we only

---

[7]To facilitate the illustration, k ranges from $0-K$.

consider the squared loss regarding the N unlabeled samples to guarantee the learning performance. We ultimately reach our objective function as

$$\min_{\mathbf{f},\mathbf{L}(\mathbf{S_0})} \sum_{i=1}^{N}(y_i - f_i)^2 + \mu \mathbf{f}^T \mathbf{L}(\mathbf{S_0})\mathbf{f} + \lambda \sum_{k=1}^{K} \left\| \frac{1}{tr(\mathbf{L}(\mathbf{S_0}))}\mathbf{L}(\mathbf{S_0}) - \frac{1}{tr(\mathbf{L}(\mathbf{S_k}))}\mathbf{L}(\mathbf{S_k}) \right\|_F^2 ,$$

where λ and μ are both nonnegative regularization parameters. To be more specific, λ penalizes the disagreement among the latent space and modalities, and μ encourages that similar popularity will be assigned to similar micro-videos.

3.6.2 OPTIMIZATION

To simplify the representation, we first define that

$$\begin{cases} \tilde{\mathbf{L}} = \frac{1}{tr(\mathbf{L}(\mathbf{S_0}))}\mathbf{L}(\mathbf{S_0}), \\ \tilde{\mathbf{L}}_k = \frac{1}{tr(\mathbf{L}(\mathbf{S_k}))}\mathbf{L}(\mathbf{S_k}). \end{cases} \tag{3.9}$$

Therefore, the objective function can be transformed to

$$\min_{\mathbf{f}} \sum_{i=1}^{N}(y_i - f_i)^2 + \lambda \sum_{k=1}^{K} \left\| \tilde{\mathbf{L}} - \tilde{\mathbf{L}}_k \right\|_F^2 + \mu \mathbf{f}^T \tilde{\mathbf{L}}\mathbf{f}, \quad \text{s.t.} \quad tr(\mathbf{L}(\mathbf{S_0})) = 1. \tag{3.10}$$

Furthermore, to optimize $\tilde{\mathbf{L}}$ more efficiently, inspired by the property that $tr(\tilde{\mathbf{L}}_k) = 1$, we let

$$\mathbf{L}(\mathbf{S_0}) = \sum_{k=1}^{K} \beta_k \tilde{\mathbf{L}}_k, \quad \text{s.t.} \quad \sum_{k=1}^{K} \beta_k = 1. \tag{3.11}$$

Consequently, we have,

$$\tilde{\mathbf{L}} = \frac{1}{tr(\mathbf{L}(\mathbf{S_0}))}\mathbf{L}(\mathbf{S_0}) = \sum_{k=1}^{K} \beta_k \tilde{\mathbf{L}}_k, \quad \text{s.t.} \quad \sum_{k=1}^{K} \beta_k = 1. \tag{3.12}$$

Interestingly, we find that β_k can be treated as the co-related degree between the latent common space and each modality. It is worth noting that we do not impose the constraint of $\beta \geq 0$, since we want to keep both positive and negative co-relations. A positive coefficient indicates the positive correlation between the modality space and the latent common space, while a negative coefficient reflects the negative correlation, which may be due to the noisy data of the modality. The larger the β_k is, the higher correlation between the latent space and the k-th modality will be. In the end, the final objective function can be written as:

$$\min_{\mathbf{f},\boldsymbol{\beta}} \sum_{i=1}^{N}(y_i - f_i)^2 + \lambda \sum_{k=1}^{K} \left\| \sum_{i=1}^{K} \beta_i \tilde{\mathbf{L}}_i - \tilde{\mathbf{L}}_k \right\|_F^2 + \mu \mathbf{f}^T \sum_{k=1}^{K} \beta_k \tilde{\mathbf{L}}_k \mathbf{f} + \theta \left\| \boldsymbol{\beta} \right\|^2,$$

$$\text{s.t.} \quad \mathbf{e}^T \boldsymbol{\beta} = 1, \tag{3.13}$$

where $\boldsymbol{\beta} = [\beta_1, \beta_2, \cdots, \beta_K]^T \in \mathbb{R}^K$ and $\mathbf{e} = [1, 1, \cdots, 1]^T \in \mathbb{R}^K$. θ is the regularization parameter, introduced to avoid the overfitting problem. We denote the objective function of Eq. (3.13)
as $\boldsymbol{\Gamma}$. We adopt the alternating optimization strategy to solve the two variables $\mathbf{f}$ and $\boldsymbol{\beta}$ in $\boldsymbol{\Gamma}$. In
particular, we optimize one variable while fixing the other one in each iteration. We keep this
iterative procedure until the $\boldsymbol{\Gamma}$ converges.

Computing β_j with f Fixed
We first fix $\mathbf{f}$ and transform the objective function $\boldsymbol{\Gamma}$ as

$$\min_{\boldsymbol{\beta}} \lambda \sum_{k=1}^{K} \sum_{t=1}^{N+M} \left\| \mathbf{M}^{(t)} \boldsymbol{\beta} - \tilde{\mathbf{l}}_k^{(t)} \right\|_F^2 + \mu \mathbf{g}^T \boldsymbol{\beta} + \theta \left\| \boldsymbol{\beta} \right\|^2, \text{s.t.} \quad \mathbf{e}^T \boldsymbol{\beta} = 1, \tag{3.14}$$

where $\mathbf{g} = [\mathbf{f}^T \tilde{\mathbf{L}}_1 \mathbf{f}, \mathbf{f}^T \tilde{\mathbf{L}}_2 \mathbf{f}, \dots, \mathbf{f}^T \tilde{\mathbf{L}}_K \mathbf{f}]^T \in \mathbb{R}^K$, $\mathbf{M}^{(t)} = [\tilde{\mathbf{l}}_1^{(t)}, \tilde{\mathbf{l}}_2^{(t)}, \dots, \tilde{\mathbf{l}}_K^{(t)}] \in \mathbb{R}^{(N+M) \times K}$ and $\tilde{\mathbf{l}}_k^{(t)} \in \mathbb{R}^{N+M}$ denotes the t-th column of $\tilde{\mathbf{L}}_k$. For simplicity, we replace $\tilde{\mathbf{l}}_K^{(t)}$ with $\tilde{\mathbf{l}}_k^{(t)} \mathbf{e}^T \boldsymbol{\beta}$, as $\mathbf{e}^T \boldsymbol{\beta} = 1$.
With the help of Lagrangian, $\boldsymbol{\Gamma}$ can be rewritten as follows:

$$\min_{\boldsymbol{\beta}} \lambda \sum_{k=1}^{K} \sum_{t=1}^{N+M} \left\| \left(\mathbf{M}^{(t)} - \tilde{\mathbf{l}}_k^{(t)} \mathbf{e}^T \right) \boldsymbol{\beta} \right\|_F^2 + \mu \mathbf{g}^T \boldsymbol{\beta} + \delta(1 - \mathbf{e}^T \boldsymbol{\beta}) + \theta \left\| \boldsymbol{\beta} \right\|^2, \tag{3.15}$$

where δ is a nonnegative Lagrange multiplier. Taking derivative of Eq. (3.15) with respect to $\boldsymbol{\beta}$,
we have

$$\frac{\partial \boldsymbol{\Gamma}}{\partial \boldsymbol{\beta}} = \mathbf{H}\boldsymbol{\beta} + \mu \mathbf{g} - \delta \mathbf{e}, \tag{3.16}$$

where

$$\mathbf{H} = 2 \left[\left(\lambda \sum_{k=1}^{K} \sum_{t=1}^{N+M} \left(\mathbf{M}^{(t)} - \tilde{\mathbf{l}}_k^{(t)} \mathbf{e}^T \right)^T \left(\mathbf{M}^{(t)} - \tilde{\mathbf{l}}_k^{(t)} \mathbf{e}^T \right) \right) + \theta \mathbf{I} \right], \tag{3.17}$$

and $\mathbf{I}$ is a $K \times K$ identity matrix. Setting Eq. (3.16) to zero, we have:

$$\boldsymbol{\beta} = \mathbf{H}^{-1}(\delta \mathbf{e} - \mu \mathbf{g}). \tag{3.18}$$

Substituting Eq. (3.18) into $\mathbf{e}^T \boldsymbol{\beta} = 1$, we have:

$$\begin{cases} \delta &= \dfrac{1 + \mu \mathbf{e}^T \mathbf{H}^{-1} \mathbf{g}}{\mathbf{e}^T \mathbf{H}^{-1} \mathbf{e}}, \\ \boldsymbol{\beta} &= \mathbf{H}^{-1} \left[\dfrac{\mathbf{e} + \mu \mathbf{e}^T \mathbf{H}^{-1} \mathbf{g} \mathbf{e}}{\mathbf{e}^T \mathbf{H}^{-1} \mathbf{e}} - \mu \mathbf{g} \right]. \end{cases} \tag{3.19}$$

According to the definition of positive-definite matrix, $\mathbf{H}$ is always positive definite and hence invertible. Therefore, $\mathbf{H}^{-1}$ is also positive definite, which ensures $\mathbf{e}^T \mathbf{H}^{-1} \mathbf{e} > 0$.

Computing f with β_j Fixed

With fixed β_j, taking derivative of $\boldsymbol{\Gamma}$ with respect to f_i, where $1 \leq i \leq N$, we have

$$\frac{\partial \boldsymbol{\Gamma}}{\partial f_i} = 2(f_i - y_i) + 2\mu \sum_{j=1}^{N+M} \tilde{L}(i, j) f_j. \tag{3.20}$$

We then take derivative of the $\boldsymbol{\Gamma}$ with respect to f_i, where $N + 1 \leq i \leq N + M$. We reach

$$\frac{\partial \boldsymbol{\Gamma}}{\partial f_i} = 2\mu \sum_{j=1}^{N+M} \tilde{L}(i, j) f_j. \tag{3.21}$$

In a vector-wise form, we restate the solution of $\mathbf{f}$ as follows:

$$\mathbf{f} = \mathbf{G}^{-1} \hat{\mathbf{y}}, \tag{3.22}$$

where $\mathbf{G} = \hat{\mathbf{I}} + \mu \sum_{k=1}^{K} \beta_k \tilde{\mathbf{L}}_k$, $\hat{\mathbf{y}} = \{y_1, y_2, \ldots, y_N, 0, 0, \ldots, 0\}$ and $\hat{\mathbf{I}} \in \mathbb{R}^{(N+M) \times (N+M)}$ is defined as follows:

$$\hat{I}(i, j) = \begin{cases} 1, & \text{if } i = j \text{ and } 1 \leq i \leq N, \\ 0, & \text{otherwise.} \end{cases} \tag{3.23}$$

3.6.3 EXPERIMENTS AND RESULTS

In this part, we conducted extensive experiments to verify our proposed TMALL model on Dataset I.

Experimental Settings

The remaining experiments were conducted over a cluster of 50 servers equipped with Intel Xeon(2x) CPU E5-2620 v3 at 2.40 GHz on 64 GB RAM, 24 cores and 64-bit Linux operating system. Regarding the deep feature extraction, we deployed Caffe framework [71] on a server equipped with a NVIDIA Titan Z GPU. The experimental results reported in this chapter were based on 10-fold cross-validation. In each round of the 10-fold cross-validation, we split Dataset I into two chunks: 90% of the micro-videos were used for training and 10% were used for testing.

We report performance in terms of normalized mean square error (nMSE) [123] between the predicted popularity and the actual popularity. The nMSE is an estimator of the overall deviations between predicted and measured values. It is defined as

$$nMSE = \frac{\sum_{i=1}(p_i - r_i)^2}{\sum_{i=1} r_i^2},$$
(3.24)

where p_i is the predicted value and r_i is the target value in ground truth.

We have three key parameters as shown in Eq. (3.10). The optimal values of these parameters were carefully tuned with the training data in each of the 10 fold. We employed the grid search strategy to obtain the optimal parameters between 10^{-5} to 10^2 with small but adaptive step sizes. In particular, the step sizes were 0.00001, 0.0001, 0.001, 0.01, 0.1, 1, and 10 for the range of [0.00001,0.0001], [0.0001,0.001], [0.001,0.01], [0.01,0.1], [0.1,1], [1,10], and [10,100], respectively. The parameters corresponding to the best nMSE were used to report the final results. For other compared systems, the procedures to tune the parameters are analogous to ensure the fair comparison. Considering one fold as an example, we observed that our model reached the optimal performance at $\lambda = 1$, $\mu = 0.01$ and $\theta = 100$.

On Model Comparison

To demonstrate the effectiveness of our proposed TMALL model, we carried out experiments on Dataset I with several state-of-the-art multi-view learning approaches.

- **Early_Fusion**. The first baseline concatenates the features extracted from the four modalities into a single joint feature vector, on which traditional machine learning models can be applied. In this work, we adopted the widely used regression model—SVR, and implemented it with the help of scikit-learn [130].

- **Late_Fusion**. The second baseline first separately predicts the popularity of micro-videos from each modality via SVR model, and then linearly integrates them to obtain the final results.

- **regMVMT**. The third baseline is the regularized multi-view learning model [190]. This model only regulates the relationships among different views within the original space.

- **MSNL**. The fourth one is the multiple social network learning (MSNL) model proposed in [149]. This model takes the source confidence and source consistency into consideration.

- **MvDA**. The fifth baseline is a multi-view discriminant analysis (MvDA) model [75], which aims to learn a single unified discriminant common space for multiple views by jointly optimizing multiple view-specific transforms, one for each view. The model exploits both the intra-view and inter-view correlations.

Table 3.1 shows the performance comparison among different models. From this table, we have the following observations. (1) TMALL outperforms the Early_Fusion and Late_Fusion. Regarding the Early_Fusion, features extracted from various sources may not fall into the same semantic space. Simply appending all features actually brings in a certain amount of noise and ambiguity. Besides, Early_Fusion may lead to the curse of dimensionality since the final feature vector would be of very high dimension. For the Late_Fusion, the fused result however might not be reasonably accurate due to two reasons. First, a single modality might not be sufficiently descriptive to represent the complex semantics of the videos. Separate results would be thus suboptimal and the integration may not result in a desired outcome. Second, it is labor-intensive to tune the fusion weights for different modalities. Even worse, the optimal parameters for one application cannot be directly applied to another one. (2) TMALL achieves better performance, as compared with regMVMT and MSNL. This could be explained that linking different modalities via a unified latent space is better than imposing disagreement penalty directly over original spaces. (3) The less satisfactory performance of MvDA indicates that it is necessary to explore the consistency among different modalities when building the latent space. (4) As compared to the multi-view learning baselines, such as regMVMT, MSNL, and MvDA, our model stably demonstrates its advantage. This signals that the proposed transductive models can achieve higher performance than inductive models under the same experimental settings. This can be explained by the fact that TMALL leverages the knowledge of testing samples.

Table 3.1: Performance comparison between our proposed TMALL model and several state-of-the-art baselines on Dataset I in terms of nMSE

Methods	nMSE	P-value
Early Fusion	59.931 ± 41.09	$9.91e\text{-}04$
Late Fusion	8.461 ± 5.34	$3.25e\text{-}03$
regMVMT	1.058 ± 0.05	$1.88e\text{-}03$
MSNL	1.098 ± 0.13	$1.42e\text{-}02$
MvDA	$0.982 \pm 7.00e\text{-}03$	$9.91e\text{-}04$
TMALL	$0.979 \pm 9.42e\text{-}03$	–

Moreover, we performed the paired t-test between TMALL and each baseline on 10-fold cross validation. We found that all the p-values are much smaller than 0.05, which shows that the performance improvements of our proposed model over other baselines are statistically significant.

On Modality Comparison

To verify the effectiveness of multi-modal integration, we also conducted experiments over different modality combinations of the four modalities. Table 3.2 summarizes the multi-modal analysis and the paired t-test results. It is obvious that the more modalities we incorporated, the better performance we can obtain. This implies the complementary relationships rather than mutual conflicting relationships among the different modalities. Moreover, we found that removing features from any of these four modalities suffers from a decrease in performance. In a sense, this is consensus with the old saying "two heads are better than one." Additionally, as the performance obtained from different combinations are not the same, this validates that incorporating β which controls the confidence of different modalities is reasonable. Interestingly, we observed that the combination without social modality obtains the worst result which indicates that the social modality plays a pivotal role in micro-video propagation, as compared to visual, textual or acoustic modality. This also validates that the features developed from social modality are much discriminative, even though they are with low-dimensions. On the other hand, the textual modality contributes the least among all modalities, as the performance of our model without textual modality still achieves good performance. This may be caused by the sparse textual description, which is usually given in one short sentence.

Table 3.2: Performance comparison among different modality combinations on Dataset I with respect to nMSE. We denote T, V, A, and S as textual, visual, acoustic, and social modality, respectively.

View Combinations	nMSE	P-value
T + V + A	$0.996 \pm 4.20e\text{-}03$	$2.62e\text{-}05$
T + A + S	$0.982 \pm 4.27e\text{-}03$	$2.59e\text{-}05$
T + V + S	$0.982 \pm 4.13e\text{-}03$	$3.05e\text{-}04$
V + A + S	$0.981 \pm 5.16e\text{-}03$	$2.16e\text{-}05$
T + V + A + S	$0.979 \pm 9.42e\text{-}03$	–

On Visual Feature Comparison

To further examine the discriminative visual features we extracted, we conducted experiments over different kinds of visual features using TMALL. We also performed significant test to validate the advantage of combining multiple features. Table 3.3 comparatively shows the performance of TMALL in terms of different visual feature configurations. It can be seen that the object, visual sentiment and aesthetic features achieve similar improvement in performance, as compared to color histogram features. This reveals that micro-videos' popularity is better reflected by their content, sentiment, and design, including what objects they contain, which emotion they convey and what design standards they follow. This is highly consistent with our oberservations and also implies that micro-videos which aim to gain high popularity need to be well designed and considered more from the visual content.

Table 3.3: Performance comparison among different visual features on Dataset I with respect to nMSE

Features	nMSE	P-value
Color Histogram	$0.996 \pm 6.88e\text{-}03$	$1.94e\text{-}04$
Object Feature	$0.994 \pm 6.71e\text{-}03$	$2.47e\text{-}04$
Visual Sentiment	$0.994 \pm 6.72e\text{-}03$	$2.49e\text{-}04$
Aesthetic Feature	$0.984 \pm 6.95e\text{-}03$	$4.44e\text{-}01$
ALL	$0.979 \pm 9.42e\text{-}03$	–

Illustrative Examples

To gain the insights of the influential factors in the task of popularity prediction of micro-videos, we comparatively illustrate a few representative examples in Figure 3.2. From this figure, we have the following observations. (1) Figure 3.2 shows three micro-video pairs. Each of the three micro-video pairs describes the similar semantics, i.e., animals, football game, and sunset, respectively, but they were published by different users. The publishers of the videos in top row are much more famous than those of the bottom. We found that the corresponding popularity of micro-videos in the second row are much lower than those in the first row, although they have no significant difference from the perspective of video contents, which clearly justifies the importance of social modality. (2) Figure 3.2 illustrates three micro-video pairs, where each pair of micro-videos were published by the same user. However, the micro-videos in the first row achieve much higher popularity than those in the second row, which demonstrates that the contents of micro-videos also contribute to their popularity. In particular, the comparisons in Figure 3.2, from left to right, are (i) the existence of "skillful pianolude" compared with "noisy dance music," (ii) "funny animals" compared with "motionless dog," and (iii) "beautiful flowers" compared with "gloomy sky." These examples indicate the necessity of developing acoustic features, visual sentiment and visual aesthetic features for the task of micro-video popularity. (3) Figure 3.2 shows a group of micro-videos, whose textual descriptions contain either superstar names, hot hashtags, or informative descriptions. These micro-videos received a lot of loops, comments, likes, and reposts. These examples thus reflect the value of textual modality.

Complexity Analysis

To theoretically analyze the computational cost of our proposed TMALL model, we first compute the complexity in the construction of $\mathbf{H}$ and $\mathbf{g}$, as well as the inverse of matrices $\mathbf{H}$ and $\mathbf{G}$.

(a) Illustration of three micro-video pairs, and each pair was published by two distinct users. The publishers of the videos in top row are much more famous than those of the bottom.

(b) Illustration of three micro-video pairs—each pair was published by the same user. The videos in the first row are much more acoustically comfortable, visually joyful, and aesthetically beautiful than those in the second row.

(c) Illustration of three popular micro-videos with different textual descriptions, which contains superstar names, hot events, and detail information, respectively.

Figure 3.2: Comparative illustration of video examples in Dataset I. They respectively: justify the importance of social, acoustic as well as visual, and textual modalities, we use three key frames to represent each video.

The construction of $\mathbf{H}$ has the time complexity of $\mathbf{O}(K^2(N + M))$. Fortunately, $\mathbf{H}$ keeps the same in each iteration, and thus can be computed by offline. The computation of $\mathbf{g}$ needs the time cost $\mathbf{O}(K(N + M)^2)$. In addition, computing the inverse of $\mathbf{H}$ and $\mathbf{G}$ has the complexity of $\mathbf{O}(K^3)$ and $\mathbf{O}((N + M)^3)$, respectively. The computation cost of $\boldsymbol{\beta}$ in Eq. (3.19) is $\mathbf{O}(K^2)$. Therefore, the speed bottleneck lies in the computation of the inverse of $\mathbf{G}$. In practice, the proposed TMALL model converges very fast, which on average takes less than 10 iterations. Overall, the learning process over 9,720 micro-videos can be accomplished within 50 s.

3.7 MULTI-MODAL TRANSDUCTIVE LOW-RANK LEARNING

Considering that we collect N samples labeled with popularity scores and M unlabeled samples, we extract K types of feature sets for this collection; hence, we obtain the feature matrix $\mathbf{X} = [\mathbf{X}_1; \mathbf{X}_2; \ldots; \mathbf{X}_K]$, where $\mathbf{X}_i \in R^{D_i \times (N+M)}$ encodes the i-th feature type and D_i is the dimensionality corresponding to the i-th feature type. Without loss of generality, we assume that columns of $\mathbf{X}_i$ are mean centered. Meanwhile, we denote $\mathbf{y} = [y_1, \ldots, y_N, 0, \ldots, 0]^T \in R^{N+M}$ as the popularity score vector for all samples, where y_i is the popularity score of the i-th sample.

In this method, we first employ four types of heterogeneous features to comprehensively characterize various aspects of micro-videos including the visual, acoustic, social, and textual modalities. As the popularity scores of micro-videos are continuous, we then formulate the task of micro-video popularity prediction as a regression problem and propose a novel low-rank multi-view learning framework named transductive low-rank multi-view regression (TL-RMVR). TLRMVR is able to learn a latent common subspace to fuse all the multi-view features such that the lowest-rank representations of the source and target are obtained for micro-video popularity prediction. Figure. 3.3 illustrates the scheme of TLRMVR, comprising of two main components: low-rank multi-view embedding learning and multi-graph regularized least squares regression. The goal of the former is to learn a set of view-specific transformation matrices by maximizing the total correlations between any two views with low-rank constraints. Due to the strength of the low-rank constraint in addressing incomplete and noisy information, low-rank representation has been successfully applied to a wide range of applications, such as subspace segmentation [95, 96, 195] and visual classification [87, 106, 197]. We are inspired to integrate the advantages of both low-rank representation and multi-view learning to enhance the robustness of feature learning. As to the second component, it is to build connections between latent low-rank representations and popularity scores. In this regard, a multi-graph regularization is constructed to improve the generalization performance and prevent overfitting. By unifying the low-rank representation and regression analysis, the lowest-rank representations shared by all views not only capture the global structure of all heterogeneous features but also indicate the regression requirements. Because the formulated objective function is non-smooth and hard to solve, we design an effective algorithm based on the augmented Lagrange multiplier (ALM) to optimize it and ensure a fast convergence.

3.7.1 OBJECTIVE FORMULATION

Low-Rank Multi-View Embedding Learning
Traditional canonical correlation analysis (CCA) [65] aims to find a common subspace in which two views of variables are fused with the maximum correlation assumption. Inspired by the success of CCA, multi-view canonical correlation analysis (MCCA) [139] was developed as a generalized CCA for multi-view scenarios. Specifically, to fully exploit the complementary properties

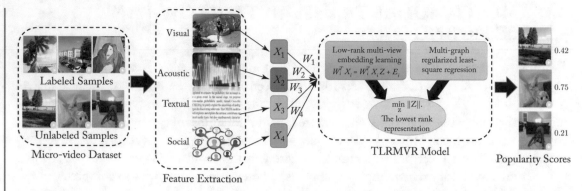

Figure 3.3: An illustration of our proposed scheme, consisting of two main components: low-rank multi-view embedding learning and multi-graph regularized least square regression.

of different views to eliminate the heterogeneity among them, MCCA attempts to find multiple basic transformation matrices $\left\{\mathbf{W}_i | \mathbf{W}_i \in R^{D_i \times D}\right\}_{i=1}^{K}$ with $1 \leq D \leq \min(D_1, D_2, \ldots, D_K)$ to, respectively, project the samples in the K views to an intrinsic low-dimensional space such that the total correlation across all view pairs is maximized while partly discarding the redundancy. Formally, it can be formulated as

$$\max_{\mathbf{W}_1, \mathbf{W}_2, \cdots, \mathbf{W}_M} \sum_{i=1}^{K} \sum_{j=1}^{K} tr\left(\mathbf{W}_i^T \mathbf{S}_{ij} \mathbf{W}_j\right) \text{ s.t. } \mathbf{W}_i^T \mathbf{S}_{ii} \mathbf{W}_i = \mathbf{I}_{D \times D}, \quad i = 1, \ldots, K, \qquad (3.25)$$

where $\mathbf{I}_{D \times D}$ is a $D \times D$ identity matrix, $\mathbf{S}_{ij} \in R^{D_i \times D_j}$ is defined as covariance matrices of $\mathbf{X}_i$ and $\mathbf{X}_j$. In a compact form, the total correlation in the common space in Eq. (3.25) can be reformulated as follows:

$$\max_{\mathbf{W}_1, \cdots, \mathbf{W}_K} tr\left(\hat{\mathbf{W}}^T \mathbf{S} \hat{\mathbf{W}}\right), \qquad \text{s.t. } \hat{\mathbf{W}}^T \hat{\mathbf{S}} \hat{\mathbf{W}} = \mathbf{I}_{D \times D}, \qquad (3.26)$$

where $\hat{\mathbf{W}} = [\mathbf{W}_1; \mathbf{W}_2; \ldots; \mathbf{W}_K] \in R^{(D_1 + D_2 + \cdots + D_K) \times D}$; $\hat{\mathbf{S}} = \text{diag}\{\mathbf{S}_{11}, \mathbf{S}_{22}, \ldots, \mathbf{S}_{KK}\} \in R^{(D_1 + \cdots + D_K) \times (D_1 + \cdots + D_K)}$ is a block-diagonal matrix; and $\mathbf{S}$ is the block matrix of the same size as $\hat{\mathbf{S}}$

$$\mathbf{S} = \begin{pmatrix} \mathbf{S}_{11} & \mathbf{S}_{12} & \ldots & \mathbf{S}_{1K} \\ \mathbf{S}_{21} & \mathbf{S}_{22} & \ldots & \mathbf{S}_{2K} \\ \ldots & \ldots & \ldots & \ldots \\ \mathbf{S}_{K1} & \mathbf{S}_{K2} & \ldots & \mathbf{S}_{KK} \end{pmatrix}.$$

The low-rank constraint is helpful for finding a more robust subspace structure and meanwhile removing the noise information from the data. Considering the aforementioned advantages, LRR assumes that the feature matrix can be decomposed into a salient part and a sparse error part. Following the scheme in LRR, we factorize the multiple view-specific transformed matrices $\mathbf{W}_1^T \mathbf{X}_1, \mathbf{W}_2^T \mathbf{X}_2, \ldots, \mathbf{W}_K^T \mathbf{X}_K$ into salient parts with a latent common low-rank structure $\mathbf{Z}$ shared by all views and their unique error matrices $\mathbf{E}_1, \mathbf{E}_2, \ldots, \mathbf{E}_K$. To better separate the salient part and the error part, we need to solve the following optimization problem:

$$\min_{\mathbf{Z}, \mathbf{E}_i} \text{rank}(\mathbf{Z}) + \lambda \sum_{i=1}^{K} \|\mathbf{E}_i\|_1 \tag{3.27}$$
$$\text{s.t. } \mathbf{W}_i^T \mathbf{X}_i = \mathbf{W}_i^T \mathbf{X}_i \mathbf{Z} + \mathbf{E}_i, \quad i = 1, \ldots, K,$$

where $\mathbf{Z} = [\mathbf{z}_1, \mathbf{z}_2, \ldots, \mathbf{z}_{N+M}] \in R^{(N+M) \times (N+M)}$ is a common low-rank representation of all samples for the view-variance structure; $\mathbf{E}_i \in R^{D \times (N+M)}$ is the unique sparse error part constrained by the $\mathcal{L}_1$-norm to handle random corruption; and $\lambda > 0$ is a balanced parameter.

There are two different understandings of the low-rank matrix $\mathbf{Z}$: one is that the matrix $\mathbf{Z}$ can be considered as an affinity matrix whose elements Z_{ij} reflect the similarity between the i-th and the j-th samples, and the other is that the feature vectors corresponding to samples of the columns of the feature matrix $\mathbf{Z}$, which plays a dominant role in representing structures learned from multiple views.

Since Eq. (3.27) is difficult to be optimized due to the non-convex rank($\cdot$), the nuclear norm $\|\mathbf{Z}\|_*$ is adopted to approximate the rank of matrix $\mathbf{Z}$. Thus, the result of Eq. (3.27) can be derived in a compact form as

$$\min_{\mathbf{Z}, \mathbf{E}} \|\mathbf{Z}\|_* + \lambda \|\mathbf{E}\|_1 \quad \text{s.t. } \hat{\mathbf{W}}^T \mathbf{X} = \hat{\mathbf{W}}^T \mathbf{X} \mathbf{Z} + \mathbf{E}, \tag{3.28}$$

where $\mathbf{E} = \left[\mathbf{E}_1^T, \mathbf{E}_2^T, \ldots, \mathbf{E}_K^T\right]^T \in R^{KD \times (N+M)}$.

Therefore, the objective function of the low-rank multi-view embedding learning can be obtained by combining the objective functions in Eqs. (3.26) and (3.28),

$$\min_{\mathbf{Z}, \mathbf{E}, \hat{\mathbf{W}}} \|\mathbf{Z}\|_* + \lambda \|\mathbf{E}\|_1 - \delta tr\left(\hat{\mathbf{W}}^T \mathbf{S} \hat{\mathbf{W}}\right) \tag{3.29}$$
$$\text{s.t. } \hat{\mathbf{W}}^T \mathbf{X} = \hat{\mathbf{W}}^T \mathbf{X} \mathbf{Z} + \mathbf{E}, \quad \hat{\mathbf{W}}^T \hat{\mathbf{S}} \hat{\mathbf{W}} = \mathbf{I},$$

where δ is a balanced parameter.

From another perspective, the core idea of the low-rank multi-view embedding learning is to utilize the relationship between features and samples. At the sample level, with the assumption that samples from different views are correlated with each other, a set of view-specific transformation matrices are obtained to project the multi-view features into a common low-dimensional subspace. At the feature level, with the greater robustness of low-rank constraint to data noise, the lowest-rank representation of each sample is obtained by revealing the underlying low-rank subspace structure spanned by the transformed samples.

Multi-Graph Regularized Least Squares Regression

As the popularity scores of micro-videos are continuous, from a narrower sense, regression analysis refers specifically to the estimation of continuous variables, which is opposite to the discrete variables used in classification. Regression analysis is widely used for predicting and forecasting tasks [54, 121, 131, 151, 156] . For example, Gelli et al. [54] modeled image popularity prediction as a $\mathcal{L}_2$ regularized $\mathcal{L}_2$ loss support vector regression (SVR) problem. Szabo et al. [151] presented a linear regression method with the maximum likelihood to predict online content popularity. Trzcinski et al. [156] introduced an SVR method with Gaussian radial basis functions (RBF) to predict the popularity of online videos. Similarity, Peng et al. [131] addressed the issue of image memorability prediction by proposing a multi-view adaptive regression model.

In this part, we use the same approach and regard the popularity prediction of micro-videos as a regression problem. For simplicity and efficiency in solving this problem, we adopt the commonly used ordinary least squares (OLS) approach, which considers a linear dependence $\mathbf{w}$ between the input feature matrix $\mathbf{Z}$ and the output popularity score $\mathbf{y}$. After adding a ridge regularization to the least squares loss part $\|\mathbf{y} - \mathbf{Z}^T \mathbf{w}\|_2^2$, we obtain a typical least squares problem with ridge regression,

$$\min_{\mathbf{w},\mathbf{Z}} \frac{1}{2} \left\| \mathbf{y} - \mathbf{Z}^T \mathbf{w} \right\|_2^2 + \alpha \left\| \mathbf{w} \right\|_2^2, \tag{3.30}$$

where $\mathbf{w} \in R^{N+M}$ is the regression coefficient and α is a parameter to balance the tradeoff between the empirical loss and the regularization penalty.

Furthermore, to provide some leeway for test samples to avoid the predicted results being regressed to zero values, we rewrite the objective function as

$$\min_{\mathbf{w},\mathbf{Z}} \frac{1}{2} \left\| \mathbf{y} - (\mathbf{Z}\mathbf{M})^T \mathbf{w} \right\|_2^2 + \alpha \left\| \mathbf{w} \right\|_2^2, \tag{3.31}$$

where $\mathbf{M}$ is a block diagonal matrix used to select labeled samples from all samples, which is defined by $\mathbf{M} = \begin{bmatrix} \mathbf{I}_{N \times N} & \mathbf{0} \\ \mathbf{0} & \mathbf{0} \end{bmatrix} \in R^{(N+M) \times (N+M)}$.

Since the low-rank embedding is learned to characterize the popularity of micro-videos, to better guide the low-rank multi-view embedding learning while avoiding the overfitting problem of test samples, we also consider the fact that the subspaces spanned by the original features and the predicted results should contain similar local geometric structures. Rather than employing a simple concatenation of features to characterize the geometric structure with a graph Laplacian, we compute a unified graph Laplacian to fuse the structures embedded in different views. Then, the geometrical structure consistency between multi-view features and the smoothness of a vectorial prediction function $f : \mathbf{z} \to R$ on graphs are preserved by minimizing the following regularizer:

$$
\begin{aligned}
\Omega\left(f\right) &= \sum_{k=1}^{K} \sum_{i,j=1}^{N+M} \left\| \mathbf{w}^T \mathbf{z}_i - \mathbf{w}^T \mathbf{z}_j \right\|_2^2 S_{ij}^k \\
&= \sum_{k=1}^{K} \mathbf{w}^T \mathbf{Z} \left(\mathbf{D}^k - \mathbf{S}^k \right) \mathbf{Z}^T \mathbf{w} \\
&= \sum_{k=1}^{K} \mathbf{w}^T \mathbf{Z} \mathbf{L}^k \mathbf{Z}^T \mathbf{w} \\
&= \mathbf{w}^T \mathbf{Z} \mathbf{L} \mathbf{Z}^T \mathbf{w},
\end{aligned}
\tag{3.32}
$$

where $\mathbf{L} = \sum_{k=1}^{K} \mathbf{L}^k$ is a unified Laplacian matrix, $\mathbf{L}^k = \mathbf{D}^k - \mathbf{S}^k$ is the graph Laplacian matrix for the k-th view, $\mathbf{S}^k$ is the weight matrix computed by the Gaussian similarity function, and $\mathbf{D}^k$ is the diagonal degree matrix with $D_{ii}^k = \sum_j S_{ij}^k$. Here, $\mathbf{S}^k$ is computed as follows:

$$
S_{ij}^k = \begin{cases} \exp\left(-\dfrac{\left\| \mathbf{x}_i^k - \mathbf{x}_j^k \right\|_2^2}{2\sigma^2} \right), & \text{if } x_i \in N_{\tilde{k}}(x_j) \text{ or } x_j \in N_{\tilde{k}}(x_i), \\ 0, & \text{otherwise,} \end{cases}
\tag{3.33}
$$

where $\mathbf{x}_i^k$ and $\mathbf{x}_j^k$ are the i-th and j-th samples in the k-th feature space, respectively; $x_i \in N_{\tilde{k}}(x_j)$ means that x_i is the $\tilde{k}$ nearest neighbor of data x_j; and σ is the radius parameter, which is simply set as the median of the Euclidean distances over all micro-video pairs.

Therefore, by combining Eqs. (3.31) and (3.32), the objective function based on low-rank representation is formulated as follows:

$$
\min_{\mathbf{w}, \mathbf{Z}} \frac{1}{2} \left\| \mathbf{y} - (\mathbf{Z}\mathbf{M})^T \mathbf{w} \right\|_2^2 + \phi \mathbf{w}^T \mathbf{Z} \mathbf{L} \mathbf{Z}^T \mathbf{w} + \alpha \left\| \mathbf{w} \right\|_2^2,
\tag{3.34}
$$

where ϕ is a balanced parameter.

To better guide the low-rank subspace learning in our previous model, we develop a quadratic term $\mathcal{G}\left(\mathbf{w}, \mathbf{Z}, \mathbf{W}\right)$ by combining supervised information (i.e., regression information and view information) and multi-graph regularizer. Based on the above formulations, the quadratic term $\mathcal{G}\left(\mathbf{w}, \mathbf{Z}, \mathbf{W}\right)$ on all samples is formulated as follows:

$$
\mathcal{G}\left(\mathbf{w}, \mathbf{Z}, \hat{\mathbf{W}}\right) = \frac{1}{2} \left\| \mathbf{y} - (\mathbf{Z}\mathbf{M})^T \mathbf{w} \right\|_2^2 + \phi \mathbf{w}^T \mathbf{Z} \mathbf{L} \mathbf{Z}^T \mathbf{w} - \delta tr\left(\hat{\mathbf{W}}^T \mathbf{S} \hat{\mathbf{W}} \right).
$$

By combining the objective functions in Eqs. (3.26) and (3.30) with Eq. (3.28), we develop our TLRMVR algorithm as follows:

$$\min_{\mathbf{w},\mathbf{Z},\mathbf{E},\hat{\mathbf{W}}} \|\mathbf{Z}\|_* + \lambda \|\mathbf{E}\|_1 + \alpha \|\mathbf{w}\|_2^2 + \beta \mathcal{G}\left(\mathbf{w},\mathbf{Z},\hat{\mathbf{W}}\right)$$

$$\text{s.t. } \hat{\mathbf{W}}^T\mathbf{X} = \hat{\mathbf{W}}^T\mathbf{X}\mathbf{Z} + \mathbf{E}, \ \hat{\mathbf{W}}\hat{\mathbf{S}}\hat{\mathbf{W}}^T = \mathbf{I}. \tag{3.35}$$

Here, we initialize $\hat{\mathbf{W}}$ by the following trace ratio equation:

$$\{\mathbf{W}_1, \mathbf{W}_2, \ldots, \mathbf{W}_K\} = \arg\max_{\mathbf{W}_1,\ldots,\mathbf{W}_K} \frac{tr\left(\hat{\mathbf{W}}^T\mathbf{S}\hat{\mathbf{W}}\right)}{tr\left(\hat{\mathbf{W}}^T\hat{\mathbf{S}}\hat{\mathbf{W}}\right)}. \tag{3.36}$$

When the low-rank representation from multi-view embedding learning and regression analysis are both performed, the lowest-rank representation shared by all views not only captures the global structure of all modalities but also indicates the regression requirements.

3.7.2   OPTIMIZATION

The objective function in Eq. (3.35) can be solved by applying the alternating direction method of multipliers (ADMM), which divides a complex problem into subproblems, where each of them is easier to handle with an iterative process. We first introduce two Lagrange multipliers $\mathbf{Y}_1$ and $\mathbf{Y}_2$ to obtain the so-called augmented Lagrangian function.

Then, we merge the last five terms into a single one: $H(\mathbf{w}, \mathbf{Z}, \mathbf{E}, \hat{\mathbf{W}}, \mathbf{Y}_1, \mathbf{Y}_2, \mu) = \beta\mathcal{G}(\mathbf{w}, \mathbf{Z}, \hat{\mathbf{W}}) + \frac{\mu}{2}\|\hat{\mathbf{W}}^T\mathbf{X} - \hat{\mathbf{W}}^T\mathbf{X}\mathbf{Z} - \mathbf{E} + \frac{\mathbf{Y}_1}{\mu}\|_F^2 + \frac{\mu}{2}\|\hat{\mathbf{W}}^T\hat{\mathbf{S}}\hat{\mathbf{W}} - \mathbf{I} + \frac{\mathbf{Y}_2}{\mu}\|_F^2$; thus, the augmented Lagrangian function of Eq. (3.35) can be reformulated as follows:

$$L\left(\mathbf{w}, \mathbf{Z}, \mathbf{E}, \hat{\mathbf{W}}, \mathbf{Y}_1, \mathbf{Y}_2, \mu\right) = \|\mathbf{Z}\|_* + \lambda \|\mathbf{E}\|_1 + \alpha \|\mathbf{w}\|_2^2 - \frac{1}{2\mu}\left(\|\mathbf{Y}_1\|_F^2 + \|\mathbf{Y}_2\|_F^2\right)$$

$$+ H\left(\mathbf{w}, \mathbf{Z}, \mathbf{E}, \hat{\mathbf{W}}, \mathbf{Y}_1, \mathbf{Y}_2, \mu\right). \tag{3.37}$$

To better interpret the process, we introduce a variable t and define $\mathbf{Z}_t, \mathbf{E}_t, \mathbf{W}_t, \mathbf{w}_t, \mathbf{Y}_{1,t}, \mathbf{Y}_{2,t}$, and μ_t as the variables updated in the t-th iteration. Under the ADMM framework, the problem L with respect to each variable in the $t + 1$ iteration is optimized as the following scheme:

For Z: We can update $\mathbf{Z}$ by dropping the terms independent of $\mathbf{Z}$ as the following scheme:

$$\mathbf{Z}_{t+1} = \arg\min_{\mathbf{Z}} \|\mathbf{Z}\|_* + H(\mathbf{Z}, \mathbf{E}, \hat{\mathbf{W}}, \mathbf{Y}_1, \mathbf{Y}_2, \mu)$$

$$= \arg\min_{\mathbf{Z}} \|\mathbf{Z}\|_* + \frac{\tau_t\mu_t}{2} \|\mathbf{Z} - \mathbf{Z}_t\|_F^2 + \langle\nabla_{\mathbf{Z}}H, \mathbf{Z} - \mathbf{Z}_t\rangle \tag{3.38}$$

$$= \arg\min_{\mathbf{Z}} \frac{1}{2} \|\mathbf{Z} - \mathbf{Z}_t + \nabla_{\mathbf{Z}}H\|_F^2 + \frac{1}{\tau_t\mu_t} \|\mathbf{Z}\|_*,$$

where $\nabla_{\mathbf{Z}} H = \beta \mathbf{w}_t (\mathbf{w}_t^T \mathbf{Z}_t \mathbf{M} - \mathbf{y}^T) \mathbf{M}^T + 2\phi \beta \mathbf{w}_t \mathbf{w}_t^T \mathbf{Z}_t \mathbf{L} - \mu_t \mathbf{X}^T \hat{\mathbf{W}}_t (\hat{\mathbf{W}}_t^T \mathbf{X} - \hat{\mathbf{W}}_t^T \mathbf{X} \mathbf{Z}_t - \mathbf{E}_t + \frac{1}{\mu_t} \mathbf{Y}_{1,t})$ is the partial derivative $H(\mathbf{Z}, \mathbf{E}, \mathbf{W}, \mathbf{Y}_1, \mathbf{Y}_2, \mu)$ with respect to $\mathbf{Z}$ and $\tau_t = \|\hat{\mathbf{W}}_t^T \mathbf{X}\|_F^2$.

The problem in Eq. (3.38) is a standard nuclear norm minimization problem, which can be approximately solved by the singular value thresholding (SVT) algorithm [14]. Specifically, suppose that the singular vector decomposition of $\mathbf{Z}_t - \nabla_{\mathbf{Z}} H$ of rank r is

$$\mathbf{Z}_t - \nabla_{\mathbf{Z}} H = \mathbf{P} \mathbf{\Sigma} \mathbf{Q}^T, \quad \mathbf{\Sigma} = \mathrm{diag}\left(\{\delta_i\}_{i=1}^r\right), \tag{3.39}$$

where $\mathbf{P}$ and $\mathbf{Q}$ are left-singular and right-singular matrices with orthogonal columns and $\mathbf{\Sigma}$ is a rectangular diagonal matrix with non-negative real numbers δ_i on the diagonal. Then, the optimal solution $\mathbf{Z}$ is $\mathbf{Z}_{t+1} = \mathcal{D}_{1/\tau_t \mu_t}(\mathbf{Z}_t - \nabla_{\mathbf{Z}} H)$. For each $1/\tau_t \mu_t \geq 0$, the soft-thresholding operator $\mathcal{D}_{1/\tau_t \mu_t}(\mathbf{Z}_t - \nabla_{\mathbf{Z}} H)$ is defined as [14]:

$$\begin{cases} \mathcal{D}_{1/\tau_t \mu_t}(\mathbf{Z}_t - \nabla_{\mathbf{Z}} H) = \mathbf{P} \mathbf{\Sigma}_{1/\tau_t \mu_t +} \mathbf{Q}^T, \\ \mathbf{\Sigma}_{1/\tau_t \mu_t +} = \mathrm{diag}(\{(\delta_i - 1/\tau_t \mu_t)_+\}), \end{cases} \tag{3.40}$$

where t_+ is the positive part of t, namely, $t_+ = \max(0, t)$.

For E: We can obtain the optimization of $\mathbf{E}$ with fixed $\mathbf{w}, \mathbf{Z}$, and $\mathbf{W}$ as follows:

$$\begin{aligned} \mathbf{E}_{t+1} &= \arg\min_{\mathbf{E}} \frac{\mu}{2} \left\| \hat{\mathbf{W}}_t^T \mathbf{X} - \hat{\mathbf{W}}_t^T \mathbf{X} \mathbf{Z}_{t+1} - \mathbf{E} \right\|_F^2 + \left\langle \mathbf{Y}_{1,t}, \hat{\mathbf{W}}_t^T \mathbf{X} - \hat{\mathbf{W}}_t^T \mathbf{X} \mathbf{Z}_{t+1} - \mathbf{E} \right\rangle + \lambda \|\mathbf{E}\|_1 \\ &= \arg\min_{\mathbf{E}} \frac{\lambda}{\mu_t} \|\mathbf{E}\|_1 + \frac{1}{2} \left\| \mathbf{E} - \hat{\mathbf{W}}_t^T \mathbf{U}_{t+1} - \mathbf{Y}_{1,t}/\mu_t \right\|_F^2, \end{aligned} \tag{3.41}$$

where $\mathbf{U}_{t+1} = \mathbf{X} - \mathbf{X} \mathbf{Z}_{t+1}$ is defined for simplicity. The optimization of Eq. (3.41) can be solved by using the shrinkage operator [90].

For w: We can optimize $\mathbf{w}$ with fixed $\mathbf{E}, \mathbf{Z}$, and $\mathbf{W}$ as follows:

$$\mathbf{w}_{t+1} = \arg\min_{\mathbf{w}} \frac{1}{2} \left\| \mathbf{y} - (\mathbf{Z}_{t+1} \mathbf{M})^T \mathbf{w} \right\|_2^2 + \phi \mathbf{w}^T \mathbf{Z} \mathbf{L} \mathbf{Z}^T \mathbf{w} + \frac{\alpha}{\beta} \|\mathbf{w}\|_2^2. \tag{3.42}$$

The above problem is actually the well-known ridge regression, whose optimal solution is $\mathbf{w}_{t+1} = \left(\mathbf{Z}_{t+1} \mathbf{M} \mathbf{M}^T \mathbf{Z}_{t+1}^T + 2\phi \mathbf{Z}_{t+1} \mathbf{L} \mathbf{Z}_{t+1}^T + \frac{2\alpha}{\beta} \mathbf{I} \right)^{-1} \mathbf{Z}_{t+1} \mathbf{M} \mathbf{y}$.

For W: By setting the derivative of L regarding $\mathbf{W}$ to zero, we have

$$\hat{\mathbf{S}} \hat{\mathbf{W}}_{t+1} (\mathbf{Y}_{2,t} + \mathbf{Y}_{2,t}^T) - 2\delta \beta \mathbf{S} \hat{\mathbf{W}}_{t+1} + \mu_t \mathbf{U}_{t+1} \mathbf{U}_{t+1}^T \hat{\mathbf{W}}_{t+1} = \mathbf{U}_{t+1} \mathbf{E}_{t+1}^T - \mathbf{U}_{t+1} \mathbf{Y}_{1,t}^T. \tag{3.43}$$

Then, $\mathbf{W}_{t+1}$ can be optimized by solving the Lyapunov equation.

Moreover, the Lagrange multipliers $\mathbf{Y}_1$ and $\mathbf{Y}_2$ are updated by the following scheme:

$$\begin{cases} \mathbf{Y}_{1,t+1} = \mathbf{Y}_{1,t} + \mu_t \left(\hat{\mathbf{W}}_{t+1}^T \mathbf{U}_{t+1} - \mathbf{E}_{t+1} \right), \\ \mathbf{Y}_{2,t+1} = \mathbf{Y}_{2,t} + \mu_t \left(\hat{\mathbf{W}}_{t+1}^T \hat{\mathbf{S}} \hat{\mathbf{W}}_{t+1} - \mathbf{I} \right). \end{cases} \tag{3.44}$$

Algorithm 3.1 Optimization of our proposed algorithm

Input: Feature matrices $\mathbf{X}$, popularity score vector $\mathbf{y}$,

parameter variables $\lambda, \alpha, \beta, \delta$.

Initialize: $\mathbf{Z}_0 = \mathbf{E}_0 = \mathbf{Y}_{1,0} = \mathbf{Y}_{2,0} = \mathbf{w} = \mathbf{0}, t = 0$,

$\phi = 1.3, \mu_0 = 10^{-6}, \mu_{max} = 10^6, t_{max} = 10^3$.

---

1. Compute the covariance matrix $\mathbf{S}_{ij}$ by $\mathbf{S}_{ij} = \mathbf{X}_i \mathbf{X}_j^T$;

2. Initialize $\hat{\mathbf{W}}_0$ by $\hat{\mathbf{W}}_0 = \arg\max_{\hat{\mathbf{W}}} \frac{tr(\hat{\mathbf{W}}^T (\mathbf{S} - \hat{\mathbf{S}}) \hat{\mathbf{W}})}{tr(\hat{\mathbf{W}}^T \hat{\mathbf{S}} \hat{\mathbf{W}})}$;

While not converged **do**

3. Fix others and update $\mathbf{Z}_{t+1}$:

$\mathbf{Z}_{t+1} = \arg\min_{\mathbf{Z}} \frac{1}{\tau\mu} \|\mathbf{Z}\|_* + \frac{1}{2} \|\mathbf{Z} - \mathbf{Z}_t + \nabla_\mathbf{Z} h\|_F^2$;

4. Fix others and update $\mathbf{E}_{t+1}$:

$\mathbf{E}_{t+1} = \arg\min_{\mathbf{E}} \frac{\lambda}{\mu_t} \|\mathbf{E}\|_1 + \frac{1}{2} \|\mathbf{E} - \hat{\mathbf{W}}_t^T \mathbf{X} + \hat{\mathbf{W}}_t^T \mathbf{X} \mathbf{Z}_{t+1} - \mathbf{Y}_{1,t}/\mu_t\|_F^2$;

5. Fix others and update $\mathbf{w}_{t+1}$:

$\mathbf{w}_{t+1} = \left(\mathbf{Z}_{t+1} \mathbf{M} \mathbf{M}^T \mathbf{Z}_{t+1}^T + 2\phi \mathbf{Z}_{t+1} \mathbf{L} \mathbf{Z}_{t+1}^T + \frac{2\alpha}{\beta} \mathbf{I}\right)^{-1} \mathbf{Z}_{t+1} \mathbf{M} \mathbf{y}$;

6. Fix others and update $\hat{\mathbf{W}}_{t+1}$:

$\hat{\mathbf{S}} \hat{\mathbf{W}}_{t+1} (\mathbf{Y}_{2,t} + \mathbf{Y}_{2,t}^T) - 2\delta\beta \mathbf{S} \hat{\mathbf{W}}_{t+1} + \mu_t \mathbf{U} \mathbf{U}^T \hat{\mathbf{W}}_{t+1} = 2\mathbf{U} \mathbf{E}_{t+1}^T - \mathbf{U} \mathbf{Y}_1^T$,

$\hat{\mathbf{W}}_{t+1} \leftarrow \text{Orthogonal}(\hat{\mathbf{W}}_{t+1})$;

7. Update the multipliers $\mathbf{Y}_{1,t+1}$ and $\mathbf{Y}_{2,t+1}$:

$\mathbf{Y}_{1,t+1} = \mathbf{Y}_{1,t} + \mu_t (\hat{\mathbf{W}}_{t+1}^T \mathbf{X} - \hat{\mathbf{W}}_{t+1}^T \mathbf{X} \mathbf{Z}_{t+1} - \mathbf{E}_{t+1})$;

$\mathbf{Y}_{2,t+1} = \mathbf{Y}_{2,t} + \mu_t (\hat{\mathbf{W}}_{t+1}^T \hat{\mathbf{S}} \hat{\mathbf{W}}_{t+1} - \mathbf{I})$;

8. Update the parameter μ_{t+1} by $\mu_{t+1} = \min(\phi\mu_t, \mu_{max})$;

9. Check the convergence conditions;

End while

Output: $\hat{\mathbf{W}}, \mathbf{E}, \mathbf{Z}, \mathbf{w}$

3.7.3   EXPERIMENTS AND RESULTS

We verify our proposed TLRMVR model over Dataset I.

Experimental Settings

As mentioned in Section 2.2, considering popularity is highly related to online social interactions, the mean values of four types of statistics, namely, the numbers of comments, reposts, likes and views/loops, are taken into account to formulate the final popularity scores of micro-videos. Figure 3.4 shows sample micro-videos that span a wide range of popularity scores.

We tested the prediction performance over 10 random splits of the Dataset I and report the average results. In each round, we used 90% of the micro-videos for training and the remaining for testing. We empirically set the adaptive parameters as $\alpha = 1$, $\delta = 0.1$, and $\lambda = 0.01$ as default. The trade-off parameters β and ϕ in TLRMVR model are selected by a grid-search approach. We first performed a coarse grid. Once we identified a ideal region, we then conducted a finer grid search on that region. Finally, we set $\phi = 0.1$ and $\beta = 0.5$.

We denote $\mathbf{X}_1, \mathbf{X}_2, \mathbf{X}_3, \mathbf{X}_4$ as the feature matrices corresponding to visual, acoustic, textual, and social views, respectively. It is worth mentioning that in this experiments, we did not consider visual quality assessment features in the visual modality.

Results and Discussions

To comprehensively validate the proposed algorithm, in the following experiments, we justified the proposed algorithm from the following six perspectives.

- **Convergence analysis**: We tested the convergence of our algorithms based on the proposed alternating algorithm.

- **Component analysis**: To verify the effectiveness of different components in our proposed scheme, we compared the prediction performance by removing each component in our method.

- **Feature analysis**: To evaluate how features contribute to the micro-video popularity prediction, we considered two forms of evaluation: (i) performance comparison among different views and (ii) performance comparison among different visual features.

- **Parameter sensitivity analysis**: We conducted experiments to investigate the influence of various weighting parameters on the prediction accuracy.

- **Comparison with state-of-the-art methods**: Performance comparisons with several state-of-the-art algorithms were conducted to demonstrate the effectiveness of our method.

Convergence Analysis   In this part, we tested the convergence of our objective function based on the proposed alternating algorithm and randomly selected a trial to report the results. Because

Figure 3.4: Micro-video examples sampled from Dataset I with various popularity scores. The micro-videos are sorted from more popular (left) to less popular (right).

$\mathbf{Z}$ is used for predicting the popularity of micro-videos, we would like to measure the variance between two sequential $\mathbf{Z}$s by the following metric:

$$D(t) = \|\mathbf{Z}_t - \mathbf{Z}_{t-1}\|_F .$$ (3.45)

This will guarantee that the final feature results will not be drastically changed. Figure 3.5 presents the absolute values of the variance during the iterations. As shown in this figure, the divergence values obtained for our proposed algorithm decrease rapidly with increasing numbers of iterations and converge after approximately 20 iterations. Based on the above analysis, the iterative criteria are essential to guarantee the convergence of our objective function. Therefore, in this chapter, we used the relative change between two consecutive iterations falling below a threshold of 1e-3 and a maximum of 30 iterations as the stopping criteria for our proposed method.

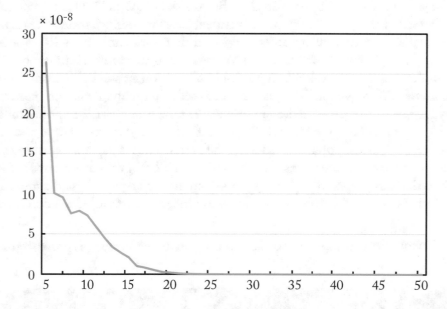

Figure 3.5: The convergence curve of our proposed TLRMVR method. The horizontal axis represents the number of iterations, and the vertical axis is the divergence between two consecutive measured $\mathbf{Z}$s.

Component Analysis   To validate the contributions of each component in our proposed framework, we compared the prediction performance by removing the relevant components.

- **noLR**: We eliminated the influence of the low-rank constraint imposed on $\mathbf{Z}$ by replacing it with the Frobenius norm.

- **noGR**: We eliminated the influence of the graph regularization term by setting $\phi = 0$.

- **noMR**: We eliminated the effect of multi-view embedding learning by setting $\delta = 0$.

- **noSP**: We eliminated the influence of supervised information, i.e., view information and regression information, by discarding both the regression coefficient and view-specific transformation matrices learning.

In the case of noSP, our algorithm degenerates to a typical unsupervised low-rank feature representation. In order to get comparable results, a least squares regression model is trained to predict popularity scores. Table 3.4 shows the prediction results of different schemes. In this table, we selected the top 50, 100, 200 images with the highest ground truth and the bottom 50, 100, 200 images with the lowest ground truth to report the average popularity scores based on their predicted results. As shown in this table, the predicted popularity scores over different ranges are Top50>Top100>Top200>Bottom200>Bottom100>Bottom50, illustrating that the behavior of the predicted results is reasonable. Moreover, we sorted the nMSE values of different methods in descending order and found that noSP>noLR>noGR = noMR; thus, the following conclusions are obtained. (1) Without supervised information, noSP performs the worst, indicating that the valuable supervised information is essential to learn a more robust prediction model. Moreover, noSP separates micro-video plurality prediction into two phases, which may lead to sub-optimal prediction results. (2) noMR and noLR impose similar significant effects on the prediction results, which means the low-rank representation and multi-view embedding learning are important in reducing the heterogeneous gap among features and alleviating the influence of feature noises. (3) Our proposed TLRMVR outperforms noGR, which demonstrates that our proposed method benefits from the use of graph regularization. This result further indicates that multi-graph regularization can indeed be employed to address multi-view

Table 3.4: Performance comparison of involved components in our proposed framework on Dataset I

	noLR	noGR	noMR	noSP	TRLMVR
Top50	0.296	0.347	0.326	0.204	0.309
Top100	0.291	0.317	0.311	0.201	0.280
Top200	0.276	0.296	0.285	0.192	0.276
Bottom200	0.253	0.271	0.269	0.172	0.265
Bottom100	0.249	0.258	0.254	0.161	0.256
Bottom50	0.246	0.251	0.251	0.157	0.249
nMSE	0.950	0.949	0.949	0.973	**0.934**
P-value	< 0.05	< 0.05	< 0.05	< 0.05	–

feature fusion problem. (4) P-value [124] is adopted to assess whether the superiority of the TL-RMVR method is statistically significant. We can discover that the P-values are smaller than the significance level of 0.05, which indicates that the null hypothesis is clearly rejected and that the improvements of TLRMVR are statistically significant.

Feature Analysis   Under our proposed framework, we investigated the influence of different features on the micro-video popularity prediction from two perspectives: (i) performance comparison of different visual-level feature combinations and (ii) performance comparison of different view-level feature combinations.

We first selected one of four visual features to represent the visual content of micro-videos and integrate with contextual, social, and acoustic cues together to conduct our experiments. Table 3.5 reports the average results over 10 random splits in terms of nMSE and P-value. From Table 3.5, we can observe the following results: (1) object features perform the best among visual features, indicating that object semantics can encode important information that makes a micro-video popular; (2) visual sentiment has a significant influence on prediction performance, illustrating that high-level sentiment semantics are helpful for micro-video popularity prediction; (3) the aesthetics exhibits better performance than the color histogram since aesthetic features specify the highly subjective nature of human perception; (4) the worst performance is still achieved by color histogram, although color histogram is effective in modeling the color perception of the human visual system; and (5) the best performance is achieved when all visual features are combined, illustrating the benefit of exploiting the complementary information offered by different visual representations.

Subsequently, we evaluated how various view-level feature combinations contribute to the popularity of micro-videos under our proposed framework. For simplicity, the features extracted from textual, visual, acoustic, and social cues are indicated as "T", "V", "A", and "S", respectively. Table 3.6 shows the average results in terms of nMSE and P-value. From Table 3.6, we can observe the following results: (1) similar to other existing studies, "T+V+A" provides the most unsatisfactory results when removing social cues, which indicates that social cues can largely facilitate popularity prediction compared to other types of cues; (2) the prediction performance of "T+A+S" sharply decreases after removing visual cues. This result shows that visual cues of micro-videos serve as an indispensable component to further improve the prediction performance; (3) "V+A+S" yields a good result of nMSE=0.955 compared to the other forms of combinations, indicating that textual cues exhibit little effect on popularity. One possible reason causing this phenomenon is that there are quite a fair number of micro-videos that lack textual descriptions. Moreover, the weak correlation between textual descriptions and micro-videos is also a common cause of this effect; and (4) when combined all view features together, the best performance is achieved with a minimum nMSE of 0.934. Additionally, it could therefore be concluded that the sequences of all cues, which are sorted in descending order in terms of their importance, is social>visual>acoustic>textual cues.

Table 3.5: Performance comparison with different visual-level feature combinations at predicting micro-video popularity on Dataset I

	Color	Object	Sentiment	Aesthetics	ALL
Top50	0.364	0.231	0.247	0.203	0.309
Top100	0.325	0.229	0.231	0.194	0.280
Top200	0.301	0.193	0.204	0.174	0.276
Bottom200	0.279	0.184	0.199	0.167	0.265
Bottom100	0.254	0.182	0.193	0.164	0.256
Bottom50	0.253	0.177	0.191	0.160	0.249
nMSE	0.975	0.967	0.969	0.971	**0.934**
P-value	< 0.05	< 0.05	< 0.05	< 0.05	–

Table 3.6: Performance comparison with different view-level feature combinations at predicting micro-video popularity on Dataset I

	T + V + A	T + A + S	T + V + S	V + A + S	TLRMVR
Top50	0.273	0.241	0.289	0.272	0.309
Top100	0.241	0.201	0.250	0.227	0.280
Top200	0.238	0.255	0.249	0.225	0.276
Bottom200	0.233	0.199	0.247	0.218	0.265
Bottom100	0.224	0.179	0.229	0.213	0.256
Bottom50	0.218	0.172	0.221	0.201	0.249
nMSE	0.979	0.970	0.958	0.955	**0.934**
P-value	< 0.05	< 0.05	< 0.05	< 0.05	–

Parameter Sensitivity Analysis

Among all the parameters in our proposed objective function, we found that the parameters ϕ and β play significant roles in affecting the prediction results. As shown in Eq. (3.42), the trade-off parameter ϕ is used to balance the effects between the graph regularization and ridge regression and the trade-off parameter β is mainly used to control the effect of the supervised loss term. Therefore, we would like to evaluate different values of ϕ and β to investigate the variation in prediction performance. In this experiment, the parameter ϕ and β are selected via a grid search in a heuristic manner, ranging from 0.05–0.30 with an interval 0.05 and ranging from 0.25–1.25 with an interval 0.25, respectively. nMSE results for various values of ϕ and β are reported in Tables 3.7 and 3.8, respectively. As shown in this table, the best performance is

Table 3.7: Performance comparison with different ϕ on our proposed framework on Dataset I

	0.05	0.10	0.15	0.20	0.25	0.30
Top50	0.370	0.309	0.283	0.238	0.230	0.198
Top100	0.347	0.280	0.269	0.227	0.219	0.187
Top200	0.330	0.276	0.251	0.212	0.205	0.175
Bottom200	0.309	0.265	0.241	0.204	0.197	0.168
Bottom100	0.298	0.256	0.231	0.196	0.189	0.162
Bottom50	0.294	0.249	0.227	0.193	0.186	0.159
nMSE	0.948	**0.934**	0.953	0.957	0.958	0.961
P-value	< 0.05	–	< 0.05	< 0.05	< 0.05	< 0.05

Table 3.8: Performance comparison with different β on our proposed framework on Dataset I

	0.25	0.50	0.75	1	1.25
Top50	0.309	0.308	0.322	0.200	0.204
Top100	0.305	0.279	0.294	0.185	0.189
Top200	0.285	0.276	0.283	0.181	0.186
Bottom200	0.263	0.265	0.268	0.175	0.181
Bottom100	0.257	0.256	0.257	0.170	0.176
Bottom50	0.252	0.249	0.251	0.166	0.172
nMSE	0.949	**0.934**	0.950	0.962	0.968
P-value	< 0.05	–	< 0.05	< 0.05	< 0.05

achieved when $\phi = 0.10$ and $\beta = 0.50$. In fact, when ϕ is set 0, our proposed method is reduced to discard the graph regularization term, which easily induces the overfitting problem. If β is set 0, our proposed method is equivalent to discard the supervised information and easily induces unsatisfactory results. This conclusions can be verified in Section 4.4.

We also evaluated the influence of various dimensions of the projection matrices. The performance of TLRMVR with different D from 10–60 is illustrated in Table 3.9. From the table, we discovered that the best dimension is 20. Too small or too large a dimension leads to a suboptimal prediction performance. It is a reasonable choice to take 20 as the reduced dimension in consideration of the complementary properties of different views.

Table 3.9: Performance comparison with different reduced dimensions D on our proposed framework on Dataset I

	10	20	30	40	50	60
Top50	0.319	0.308	0.318	0.316	0.316	0.297
Top100	0.288	0.279	0.286	0.277	0.277	0.270
Top200	0.274	0.276	0.275	0.274	0.274	0.262
Bottom200	0.269	0.265	0.269	0.267	0.267	0.256
Bottom100	0.250	0.256	0.252	0.249	0.249	0.245
Bottom50	0.243	0.249	0.243	0.241	0.241	0.236
nMSE	0.950	**0.934**	0.947	0.949	0.951	0.953
P-value	< 0.05	–	< 0.05	< 0.05	< 0.05	< 0.05

Comparison with state-of-the-art methods   We compared our proposed scheme with several existing state-of-the-art methods, including multiple linear regression (MLR), lasso regression, support vector regression (SVR) [147], RegMVMT [190], multi-feature learning via hierarchical regression (MLHR) [181], multiple social network learning (MSNL) [149], multi-view discriminant analysis [75], transductive multi-modal learning (TMALL) [24], and extreme learning machine (ELM) [67].

- **MLR**: Multiple linear regression (MLR) attempts to capture the dependency between two or more independent variables and a response variable using a linear equation, which is an extension of classical linear regression.

- **Lasso**: Lasso regression considers both variable selection and regularization to enhance the prediction performance.

- **SVR**: Support vector regression [147] is a classical regression technique with a maximum margin criterion. We combined all the features together with an RBF kernel to learn a non-linear SVR in a high-dimensional kernel-induced feature space.

- **RegMVMT**: RegMVMT [190] is an inductive learning framework to address the general multi-view learning problem, in which the co-regularization technique is utilized to enforce the agreement with other views on unlabeled samples.

- **MLHR**: The multi-feature fusion via hierarchical regression [181] is a semi-supervised learning method, which has been developed to explore the structural information embedded in data from the view of multi-feature fusion.

- **MSNL**: Multiple social network learning (MSNL) [149] is proposed to address the incomplete data in source confidence and source consistency by modeling source confidence and source consistency simultaneously.

- **MvDA**: Multi-view discriminant analysis (MvDA) [75] is a multi-view learning model, which has been developed to search for a latent common space by enforcing the view-consistency of multi-linear transforms.

- **TMALL**: The transductive multi-modal learning (TMALL) model is presented for predicting the popularity of micro-videos, in which different modal features can be unified and preserved in a latent common space to address the insufficient information problems.

- **ELM**: As ELM [68, 154] can embed a wide type of feature mappings, Huang et al. [67] extended ELM to kernel learning and proposed a unified learning mechanism for regression applications with higher scalability and less computational complexity.

Table 3.10 reports the prediction performances of our proposed method and other state-of-the-art algorithms. From this table, we have the following observations: (1) our proposed TL-RMVR performs the best among all the comparative methods; (2) lasso and MLR performs the worst, as expected, indicating that simple feature selection and linear regression are insufficient to predict the popularity of micro-videos; (3) in contrast to Lasso and MLR, the algorithms, including RegMVMT, MLHR, MSNL, MvDA, and TMALL, also perform comparably, which can be attributed to their ability to solve the multi-view/modal feature fusion problem; (4) after employing the RBF kernel to deal with multiple features, the SVR model provides a significant

Table 3.10: Performance comparison between our proposed method and several state-of-the-art methods on Dataset I

Methods	nMSE	P-value
MLR	$1.442 \pm 2.55e\text{-}01$	$1.05e\text{-}07$
Lasso	$1.568 \pm 1.72e\text{-}01$	$4.42e\text{-}08$
SVR	$0.991 \pm 5.00e\text{-}02$	$7.36e\text{-}06$
RegMVMT	$1.058 \pm 4.33e\text{-}05$	$1.88e\text{-}03$
MLHR	$1.167 \pm 1.40e\text{-}02$	$4.75e\text{-}06$
MSNL	$1.098 \pm 1.30e\text{-}01$	$2.11e\text{-}04$
MvDA	$0.982 \pm 7.00e\text{-}03$	$2.62e\text{-}05$
TMALL	$0.979 \pm 9.42e\text{-}03$	$1.43e\text{-}08$
ELM	$0.982 \pm 6.68e\text{-}05$	$3.71e\text{-}07$
TLRMVR	$\mathbf{0.934 \pm 7.67e\text{-}04}$	–

improvement in the micro-video popularity prediction tasks; (5) as stated in [67], SVR provides a suboptimal learning solution compared to ELM. Accordingly, the results present that ELM achieves better prediction performance than SVR; and (6) although MSNL and TMALL are appropriate to deal with incomplete data, TLRMVR still outperforms them, thus demonstrating the effectiveness of our approach.

Complexity Discussion   In order to analyze the complexity of TLRMVR, we suppose that the number of samples is larger than the dimension of data, i.e., $(N + M) > (D_1 + D_2 + \cdots + D_K)$. As discussed previously, we can find that the main computational complexity comes from the following parts.

- nuclear norm calculation in step 3,

- matrix inverse calculation in step 5, and

- solving the Lyapunov equation in step 6.

The computational complexity of nuclear norm is at most $O((N + M)^3)$. The matrix inverse costs $O((N + M)^3)$. The typical cost of the Lyapunov equation needs $O((N + M)^3)$. If the algorithm converges within T iteration steps for its outer loop, the upper bound of the complexity is $O(3T(N + M)^3)$. The simulations of our proposed algorithm are carried out in MATLAB 7.0.1 environment running in Core 3 Quad, 3.6-GHZ CPU with 8-GB RAM. The learning and testing processes over all micro-videos can be accomplished within 1,627 s. The speed bottleneck lies in the number of samples. Therefore, to handle large-scale dataset, Coppersmith and Winograd [32] presented a new method to accelerate matrix inversion to $O((N + M)^{2.376})$. Liu et al. [94] offered a more efficient method to solve nuclear norm calculation.

3.8   SUMMARY

In this chapter, we first present a novel transductive multi-modal learning method (TMALL), to predict the popularity of micro-videos. In particular, TMALL works by learning an optimal latent common space from multi-modalities of the given micro-videos, in which the popularity of micro-videos are much more distinguishable. The latent common space is capable of unifying and preserving information from different modalities, and it helps to alleviate the modality limitation problem. To verify our model, we built a benchmark dataset and extracted a rich set of popularity-oriented features to characterize micro-videos from multiple perspectives. By conducting extensive experiments, we draw the following conclusions: (1) the optimal latent common space exists and works; (2) the more modalities we incorporate to learn the common space, the more discriminant it is; and (3) the features extracted to describe the social and content influence are representative.

Also, we introduce a novel low-rank multi-view embedding framework to alleviate the heterogeneous, interconnected, and noisy problems in micro-video popularity prediction. By taking advantages of low-rank representation and multi-view learning, we effectively integrated all heterogeneous features extracted from different views into a common feature subspace and achieved enhanced robust feature representation for regression analysis. We also designed an effective optimization algorithm to solve the proposed model.

CHAPTER 4

Multimodal Cooperative Learning for Micro-Video Venue Categorization

4.1 BACKGROUND

As is known, geographic information benefits many services, such as location-based search, recommendation, and social networking. However, in real-world scenarios, few users tag their micro-videos with specific geographic information due to privacy concerns. Specifically, as reported in [192], around 98.78% micro-videos do not have location information. Despite its significance, we have to mention that it is hard, if not impossible, to infer the specific location information, such as "American Airlines Arena in Florida, USA." Instead, we turn to infer the venue category of a given micro-video, such as "Basketball Court." And technically speaking, venue category estimation of micro-videos can be treated as a multi-modal fusion problem and solved by integrating the geographic cues from visual, acoustic, and textual modalities of micro-videos. Motivated by this, in this chapter, we propose three different multi-modal learning models to infer the venue information of micro-videos.

4.2 RESEARCH PROBLEMS

Inferring the venue categories from micro-videos is non-trivial, due to the following challenges.

(1) Heterogeneous multi-modalities. Similar to the traditional long videos, like the ones in YouTube, micro-videos are also the unity of textual, visual, and acoustic modalities, which characterizes the video content from multiple complementary views. Although some efforts have been dedicated to data fusion [103, 116, 148], how to model the relatedness among multi-modalities and effectively fuse them is still an open research question.

(2) Sparse information. The most prominent attribute of micro-video platforms is that they are thriving heavily in the realm of shortness and instant. For example, Vine allows users to upload about 6 s videos online; Snapchat offers its users the option to create 10 s micro-videos; and Viddy limits the length of its upload videos to 30 s. Persuasively, short length makes video production and broadcasting easy, downloading timely, and playing fluent on portable devices, however, in contrast to the traditional long videos, micro-videos are comparatively short, thereby

merely able to convey only one or just a few high-level themes or concepts. Consequently, it is necessary to learn the high-level and sparse representations of micro-videos.

(3) Low-quality. Most portable devices have nothing to offer for video stabilization. Some videos can thus be shaky or bumpy, which greatly hinders the visual expression. Furthermore, the audio track that comes along with the video, can be in different types of distortion and noise, such as buzzing, hums, hisses, and whistling, which is probably caused by the poor microphones or complex surrounding environments.

(4) Information loss. Apart from acoustic and visual modalities, micro-videos are, more often than not, uploaded with textual descriptions, which express some useful cues that may be not available in the other two modalities. However, the textual information may be not well correlated with visual and acoustic cues. Moreover, according to our statistics upon 276,624 Vine videos, more than 11.4% of them do not have such text, probably the results of users' casual habits. This serious information missing problem greatly reduces the usability of textual modality.

(5) Hierarchical structure. The venues of micro-videos are organized into hundreds of categories, which are not independent but hierarchically correlated. Part of this structure is shown in Figure 1.2. How to explore such structure to guide the venue category estimation is largely untapped.

Moreover, when organizing the micro-videos, we have to consider one indispensable factor, i.e., online learning. On the one hand, micro-videos are often easily shot and instantly shared at the mobile end, timeliness is therefore one of their highlights. In light of this, efficient online operations deserve our attention. On the other hand, according to our statistics over 2 million Vine videos, only 1.22% of them have been labeled with venue information and the tree structure of the venue categories holds 821 leaf nodes. Thereby, it is hard to acquire sufficient training samples to build a robust model for the micro-video categorization. Fortunately, micro-videos are continuously uploaded and we expect to incrementally strengthen our model by leveraging the knowledge of incoming samples.

To address the aforementioned challenges, in this chapter we develop three schemes from different perspectives in order to organize micro-videos into a tree taxonomy.

4.3   RELATED WORK

Our work is related to a broad spectrum of multimedia location estimation, multi-modal multi-task learning, and dictionary learning.

4.3.1   MULTIMEDIA VENUE ESTIMATION

Nowadays, it has become convenient to capture images and videos on the mobile end and associate them with GPS tags. Such a hybrid data structure can benefit a wide variety of potential multimedia applications, such as location recognition [58], landmark search [23], augmented reality[15], and commercial recommendations [183]. It hence has attracted great attention from the multimedia community. Generally speaking, prior efforts can be divided into two categories: mono-modal venue estimation [15, 23, 27] and multi-modal venue estimation [29, 50, 58]. Approaches in the former category extract a rich set of visual features from images and leverage the visual features to train either shallow or deep models to estimate the venues of the given images. As reported in [58], the landmark identification [27] and scene classification [15] of images are the key factors to recognize the venues. The basic philosophy behind these approaches is that certain visual features in images correlate strongly with certain geographies even if the relationship is not strong enough to specifically pinpoint a location coordinate. Beyond the mono-modal venue estimation which only takes the visual information into consideration, multi-modal venue estimation works by inferring the geo-coordinates of the recording places of the given videos by fusing the textual metadata and visual or acoustic cues [29, 49]. Friendland et al. [49] determined the geo-coordinates of the Flickr videos based on both textual metadata and visual cues. Audio tracks from the Placing Task 2011 dataset videos were also used to train a location estimation models and it achieved reasonable performance [84]. The main idea is that the integration of multiple modalities can lead to better results, and it is consistent to the old saying "two heads are better than one." However, multi-modal venue estimation is still at its infant stage, and more efforts should be dedicated to improve this line of research.

Noticeably, the venue granularity of the targeted multimedia entities in the aforementioned literature varies significantly. Roughly, the spatial resolutions are in three levels: city-level [50, 58], within-city-level [23, 89, 143], and close-to-exact GPS level [49]. City-level and within-city-level location estimation can be applied to multimedia organization [33], location visualization [23], and image classification [153]. However, their granularities are large, which may be not suitable for some application scenarios, such as business venue discovery [21]. The granularity of close-to-exact GPS level is finer; nevertheless, it is hard to estimate the precise coordinates, especially for the indoor cases. For example, it is challenging to distinguish an office on the third floor and a coffee shop on the second floor within the same building, since the GPS is not available indoors.

Our work differs from the above methods in the following two aspects: (1) we focus on the estimation of venue category which is neither city-level nor the precise location. This is because venue category is more of an abstract concept than single venue name, which can help many applications for personalized and location-based services/marketing [21]; and (2) micro-videos are the medium between images and traditional long videos, which pose tough challenges.

4.3.2   MULTI-MODAL MULTI-TASK LEARNING

The literature on the multi-task problem with multi-modal data is relatively sparse. He et al. [59] proposed a graph-based iterative framework for multi-view multi-task learning (i.e., *IteM*^{2}) and applied it to text classification. However, it can only deal with problems with non-negative feature values. In addition, it is a transductive model. Hence, it is unable to generate predictive models for independent and unseen testing samples. To address the intrinsic limitations of transductive models, Zhang et al. [190] proposed an inductive multi-view multi-task learning model (i.e., regMVMT). It penalizes the disagreement of models learned from different sources over the unlabeled samples. However, without prior knowledge, simply restricting all the tasks to be similar is inappropriate. As an extension of regMVMT, an inductive convex shared structure learning algorithm for multi-view multi-task problem (i.e., CSL-MTMV) was developed in [72]. Compared to regMVMT, CSL-MTMV considers the shared predictive structure among multiple tasks.

However, none of the methods mentioned above can be applied to venue category estimation directly. This is due to the following reasons: (1) $IteM^2$, regMVMT, and CSL-MTMV are all binary classification models, of which the extension to multi-class or regression problem is nontrivial, especially when the number of classes is large; and (2) the tasks in venue category prediction are pre-defined as a hierarchical structure.

4.3.3   DICTIONARY LEARNING

Dictionary learning [126, 193] is a representation learning method, aiming to learn an overcomplete dictionary in which only a few atoms can be linearly combined to well approximate a given data sample [81]. Roughly speaking, we can group the existing efforts into two categories: unsupervised and supervised dictionary learning. The main concern of the former one is to reconstruct the original data as accurate as possible via minimizing the reconstruction error. They achieve expected performance in reconstruction tasks, such as denoising [46], inpainting [110], restoring [179], and coding [109]. They, however, may lead to suboptimal performance in the classification tasks [97, 180], wherein the ultimate goal is to make the learned dictionary and corresponding sparse representation as discriminative as possible [108]. This motivates the emergence of supervised dictionary learning [111, 160], which leverages the class labels in the training set to build a more discriminative dictionary for the particular classification task at hand. They have been well adapted to many applications with better performance, such as visual tracking [174], recognition [73], event detection [178], retrieval [172], classification [6], image super-resolution, and photo-sketch synthesis [165]. Regardless of whether it is unsupervised or not, the existing dictionary learning methods are mostly based on a single modality, and few of them encode the hierarchical data structure into the dictionary learning.

4.4   MULTIMODAL CONSISTENT LEARNING

To intuitively demonstrate our proposed model, we first introduce two assumptions.

1. **Multi-modal consistency.** We assume that there exists a common discriminative space for micro-videos, originating from their multimodalities. Micro-videos can be comprehensively described in this common space and the venue categories are more distinguishable in this space. The space over each individual modality can be mathematically mapped to the common space with a small difference.

2. **Hierarchical Structure.** The tasks (venue categories) are organized into a tree structure. We assume that such structure encodes the relatedness among tasks and leveraging this prior knowledge is able to boost the learning performance.

Based on these assumptions, we introduce our first model for micro-video venue categorization, which is a TRee-guided mUlti-task Multi-modal leArNiNg model, TRUMANN for short. As illustrated in Figure 4.1, this model intelligently learns a common feature space from multi-modal heterogeneous spaces and utilizes the learned common space to represent each micro-video. Meanwhile, the TRUMANN treats each venue category as a task and leverages the pre-defined hierarchical structure of venue categories to regularize the relatedness among tasks via a novel group lasso. These two objectives are accomplished within a unified framework. As a byproduct, the tree-guided group lasso is capable of learning task-sharing and task-specific features.

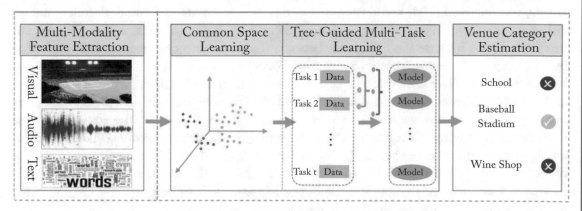

Figure 4.1: Graphical representation of our TRUMANN framework.

Formally, suppose we have a set of N micro-video samples. Each has S modalities and is associated with one of T venue categories. In this work, we treat each venue category as a task. We utilize $\mathbf{X}^s = [\mathbf{x}_1^s, \mathbf{x}_2^s, \ldots, \mathbf{x}_N^s] \in \mathbb{R}^{N \times D^s}$ to denote the representation of N samples with a D^s dimensional feature space from the s-th modality, and utilize $\mathbf{Y} = [\mathbf{y}_1, \mathbf{y}_2, \ldots, \mathbf{y}_N]^T \in \mathbb{R}^{N \times T}$ to denote the labels of the N samples over the T pre-defined tasks $\{t_1, t_2, \ldots, t_T\}$. Our objective is to jointly learn the mapping matrix $\mathbf{A}^s$ from the individual space $\mathbf{X}^s$ to the common space $\mathbf{B} \in \mathbb{R}^{N \times K}$, and learn the optimal coefficient matrix $\mathbf{W} = [\mathbf{w}_1, \mathbf{w}_2, \ldots, \mathbf{w}_T] \in \mathbb{R}^{K \times T}$. Based on $\mathbf{A}^s$ and $\mathbf{W}$ we are able to estimate the venue categories for the unseen videos. In the following, we will detail each process in a stepwise way.

Objective Formulation   Common space learning [27, 56, 168] over multiple modalities or views has been well studied. Theoretically, it can capture the intrinsic and latent structure of data, which preserves information from multiple modalities. It is thus able to alleviate the fusion and disagreement problems of the classification tasks over multiple modalities [56]. Based upon our first assumption, we propose a joint optimization framework which minimizes the reconstruction errors over multiple modalities of the data, and avoids overfitting using Frobenius norm on the transformation matrices. It is formally defined as

$$\min_{\mathbf{A}^s, \mathbf{B}} \frac{\lambda_1}{2} \sum_{s=1}^{S} \left\| \mathbf{X}^s \mathbf{A}^s - \mathbf{B} \right\|_F^2 + \frac{\lambda_2}{2} \sum_{s=1}^{S} \left\| \mathbf{A}^s \right\|_F^2 , \tag{4.1}$$

where $\mathbf{B} \in \mathbb{R}^{N \times K}$ is the representation matrix in the common space learned from all modalities, and K is the latent feature dimension. $\mathbf{A}^s \in \mathbb{R}^{D_s \times K}$ is the transformation matrix from the original feature space over the s-th modality to the common space; and λ_1 and λ_2 are nonnegative tradeoff parameters.

Hierarchical Multi-Task Learning   Although the existing multi-task learning methods, such as graph-regularized [204] and clustering-based [70], achieve sound theoretical underpinnings and great practical success, the tree-guided method [79] is more suitable and feasible for our problem. This is because the relatedness between the venue categories are naturally organized into a hierarchical tree structure by experts from Foursquare. As Figure 1.2 shows, the relatedness among different tasks can be characterized by a tree τ with a set of nodes $\mathcal{V}$, where the leaf nodes and internal nodes represent tasks and groups of the tasks, respectively. Intuitively, each node $v \in \mathcal{V}$ of the tree can be associated with a corresponding group $\mathcal{G}_v = \{t_i\}$, which consists of all the leaf nodes t_i belonging to the subtree rooted at the node v. To capture the strength of relatedness among tasks within the same group $\mathcal{G}_v$, we assign a weight e_v to node v according to an affinity function, which will be detailed in the next part. Moreover, the higher level the internal node locates at, the weaker relatedness it controls, and hence the smaller weight it obtains. Therefore, we can formulate such tree-guided multi-task learning as follows:

$$\min_{\mathbf{W}, \mathbf{B}} \Gamma = \frac{1}{2} \left\| \mathbf{Y} - \mathbf{B} \mathbf{W} \right\|_F^2 + \frac{\lambda_3}{2} \sum_{v \in \mathcal{V}} e_v \left\| \mathbf{W}_{\mathcal{G}_v} \right\|_{2,1} , \tag{4.2}$$

where $\mathbf{W}_{\mathcal{G}_v} = \{\mathbf{w}_i : t_i \in \mathcal{G}_v\} \in \mathbb{R}^{K \times |\mathcal{G}_v|}$ is the coefficient matrix of all the leaf nodes rooted at v, where each column vector is selected from $\mathbf{W}$ according to the members within the task group $\mathcal{G}_v$; $\|\mathbf{W}_{\mathcal{G}_v}\|_{2,1} = \sum_{k=1}^{K} \sqrt{\sum_{t_i \in \mathcal{G}_v} w_{ki}^2}$ is the $\ell_{2,1}$-norm regularization (i.e., group lasso) which is capable of selecting features based on their strengths over the selected tasks within the group $\mathcal{G}_v$, and in this way, we can simultaneously learn the task-sharing features and task-specific features. Lastly, the nonnegative parameter λ_3 regulates the sparsity of the solution regarding $\mathbf{W}$.

By integrating the common space learning function in Eq. (4.1) and the tree-guided multi-task learning framework in Eq. (4.2), we reach the final objective function as follows:

$$\min_{\mathbf{W}, \mathbf{B}} \Gamma = \frac{1}{2} \left\| \mathbf{Y} - \mathbf{B} \mathbf{W} \right\|_F^2 + \frac{\lambda_1}{2} \sum_{s=1}^{S} \left\| \mathbf{X}^s \mathbf{A}^s - \mathbf{B} \right\|_F^2 + \frac{\lambda_2}{2} \sum_{s=1}^{S} \left\| \mathbf{A}^s \right\|_F^2 + \frac{\lambda_3}{2} \sum_{v \in \mathcal{V}} e_v \left\| \mathbf{W}_{\mathcal{G}_v} \right\|_{2,1} .$$

$$\tag{4.3}$$

4.4.1 OPTIMIZATION

We adopt the alternating optimization strategy to solve the three variables $\mathbf{A}^s$, $\mathbf{B}$, and $\mathbf{W}$ in Eq. (4.3). To be more specific, we optimize one variable while fixing the others in each iteration. We keep this iterative procedure until the objective function converges.

Computing $\mathbf{A}^s$ with $\mathbf{B}$ and $\mathbf{W}$ fixed: we first fix $\mathbf{B}$ and $\mathbf{W}$, and take derivative of Γ with respect to $\mathbf{A}^s$. We have

$$\frac{\partial \Gamma}{\partial \mathbf{A}^s} = \lambda_1 \left(\mathbf{X}^s \mathbf{A}^s - \mathbf{B} \right) \mathbf{X}^s + \lambda_2 \mathbf{A}^s. \tag{4.4}$$

By setting Eq. (4.4) to zero, it can be derived that

$$\mathbf{A}^s = \left(\lambda_1 \mathbf{X}^{sT} + \lambda_2 \mathbf{I} \right)^{-1} \left(\lambda_1 \mathbf{X}^{sT} \mathbf{B} \right), \tag{4.5}$$

where $\mathbf{I} \in \mathbb{R}^{D_s \times D_s}$ is an identity matrix. The first term in Eq. (4.5) can be easily proven to be positive definite and hence invertible according to the definition of positive-definite matrix.

Computing $\mathbf{B}$ with $\mathbf{A}^s$ and $\mathbf{W}$ fixed: with $\mathbf{A}^s$ and $\mathbf{W}$ fixed, we compute the derivative of Γ regarding $\mathbf{B}$ as follows:

$$\frac{\partial \Gamma}{\partial \mathbf{B}} = \lambda_1 \sum_{s=1}^{S} \left(\mathbf{B} - \mathbf{X}^s \mathbf{A}^s \right) + \left(\mathbf{B} \mathbf{W} \mathbf{W}^T - \mathbf{Y} \mathbf{W}^T \right). \tag{4.6}$$

By setting Eq. (4.6) to zero, we have

$$\mathbf{B} = \left(\mathbf{Y} \mathbf{W}^T + \lambda_1 \sum_{s=1}^{S} \mathbf{X}^s \mathbf{A}^s \right) \left(\lambda_1 S \mathbf{I} + \mathbf{W} \mathbf{W}^T \right)^{-1}, \tag{4.7}$$

where $\lambda_1 S \mathbf{I} + \mathbf{W} \mathbf{W}^T$ can be easily proven to be invertible according to the definition of positive-definite matrix.

Computing $\mathbf{W}$ with $\mathbf{A}^s$ and $\mathbf{B}$ fixed: Considering that the last term in Eq. (4.3) is not differentiable, we use an equivalent formulation of it, which has been proven by [3], to facilitate the optimization as follows:

$$\frac{\lambda_3}{2} \left(\sum_{v \in \mathcal{V}} \| \mathbf{W}_{\mathcal{G}_v} \| \right)^2. \tag{4.8}$$

Still, it is intractable. We thus further resort to another variational formulation of Eq. (4.8). According to the Cauchy-Schwarz inequality, given an arbitrary vector $\mathbf{b} \in \mathbb{R}^M$ such that $\mathbf{b} \neq 0$, we have

$$\sum_{i=1}^{M} |b_i| = \sum_{i=1}^{M} \theta_i^{\frac{1}{2}} \theta_i^{-\frac{1}{2}} |b_i| \leq \left(\sum_{i=1}^{M} \theta_i \right)^{\frac{1}{2}} \left(\sum_{i=1}^{M} \theta_i^{-1} b_i^2 \right)^{\frac{1}{2}} \leq \left(\sum_{i=1}^{M} \theta_i^{-1} b_i^2 \right)^{\frac{1}{2}}, \tag{4.9}$$

where θ_i's are the introduced variables that should satisfy $\sum_{i=1}^{M} \theta_i = 1$, $\theta_i > 0$ and the equality holds for $\theta_i = |b_i|/\|\mathbf{b}\|_1$. Based on this preliminary, we can derive the following inequality:

$$\left(\sum_{v \in \mathcal{V}} e_v \|\mathbf{W}_{\mathcal{G}_v}\|\right)^2 \leq \sum_{k=1}^{K} \sum_{v \in \mathcal{V}} \frac{e_v^2 \|\mathbf{w}_{\mathcal{G}_v}^k\|_2^2}{q_{k,v}}, \tag{4.10}$$

where $\sum_k \sum_v q_{k,v} = 1$, $q_{k,v} \geq 0$, $\forall k, v$; $\mathbf{w}_{\mathcal{G}_v}^k$ denotes the k-th row vector of the group matrix $\mathbf{W}_{\mathcal{G}_v}$. It worth noting that the equality holds when

$$q_{k,v} = \frac{e_v^2 \|\mathbf{w}_{\mathcal{G}_v}^k\|_2^2}{\sum_{k=1}^{K} \sum_{v \in \mathcal{V}} e_v^2 \|\mathbf{w}_{\mathcal{G}_v}^k\|_2^2}. \tag{4.11}$$

Thus far, we have theoretically derived that minimizing γ with respect to $\mathbf{W}$ is equivalent to minimizing the following convex objective function:

$$\min_{\mathbf{W}, q_{k,v}} \frac{1}{2} \|\mathbf{Y} - \mathbf{BW}\|_F^2 + \frac{\lambda_1}{2} \sum_{s=1}^{S} \|\mathbf{X}^s \mathbf{A}^s - \mathbf{B}\|_F^2 + \frac{\lambda_2}{2} \sum_{s=1}^{S} \|\mathbf{A}^s\|_F^2 + \frac{\lambda_3}{2} \sum_{k=1}^{K} \sum_{v \in \mathcal{V}} \frac{\left\|e_v \mathbf{W}_{\mathcal{G}_v}^k\right\|^2}{q_{k,v}}. \tag{4.12}$$

To facilitate the computation of the derivative of objective function Γ with respect to $\mathbf{w}_t$ for the t-th task, we define a diagonal matrix $\mathbf{Q}^t \in \mathbb{R}^{K \times K}$ with the diagonal entry as follows:

$$\mathbf{Q}_{kk}^t = \sum_{\{v \in \mathcal{V} | t \in v\}} \frac{e_v^2}{q_{k,v}}. \tag{4.13}$$

We ultimately have the following objective function:

$$\min_{\mathbf{W}, \mathbf{Q}} \sum_{t=1}^{T} \|\mathbf{y}_t - \mathbf{Bw}_t\|_F^2 + \frac{\lambda_1}{2} \sum_{s=1}^{S} \|\mathbf{X}^s \mathbf{A}^s - \mathbf{B}\|_F^2 + \frac{\lambda_2}{2} \sum_{s=1}^{S} \|\mathbf{A}^s\|_F^2 + \frac{\lambda_3}{2} \sum_{t=1}^{T} \mathbf{w}_t^T \mathbf{Q}^t \mathbf{w}_t. \tag{4.14}$$

The alternative optimization strategy is also applicable here. By fixing $\mathbf{Q}^t$, taking derivative of the above formulation regarding $\mathbf{w}_t$, and setting it to zero, we reach

$$\mathbf{w}_t = \left(\mathbf{B}^T \mathbf{B} + \lambda_3 \mathbf{Q}^t\right)^{-1} \left(\mathbf{B}^T \mathbf{y}_t\right). \tag{4.15}$$

Once we obtain all the $\mathbf{w}_t$, we can easily compute $\mathbf{Q}^t$ based on Eq. (4.11).

4.4.2 TASK RELATEDNESS ESTIMATION

According to our assumption, the hierarchical tree structure of venue categories plays a pivotal role to boost the learning performance in our model. Hence, the key issue is how to precisely characterize and model the task relatedness in the tree, namely, how to estimate the reasonable weight e_v for each node v in the tree appropriately. Although the existing tree-guided multi-task learning approaches [57, 150] have addressed this issue by exploring the geometric structure, they do not consider the semantic relatedness among tasks. To remedy this problem, we aim to model the intrinsic task relatedness based on the feature space. Toward this goal, we introduce the affinity measurement of the node group proposed in [98]. A high affinity value e_v of the node group $\mathcal{G}_v$ indicates the dense connections and compact relations among the leaf nodes within the given group. We hence can employ the affinity measurement to characterize the task relatedness in the tree.

To facilitate the affinity measurement of each node group $\mathcal{G}_v$, we need to obtain the pair-wise similarity between all leaf nodes. For simplicity, we utilize the adjacency matrix $\mathbf{S} \in \mathbb{R}^{T \times T}$ to denote the pairwise similarity matrix and the entry S_{ij} to capture the non-negative relatedness between the i-th and j-th leaf nodes, which can be formulated as

$$S_{ij} = \exp\left(-\frac{\|\bar{\mathbf{x}}_i - \bar{\mathbf{x}}_j\|^2}{\theta^2}\right), \tag{4.16}$$

where $\bar{\mathbf{x}}_i$ represents the mean feature vector of the samples belonging to the i-th venue category which can be extracted from the training dataset; θ is radius parameter that is simply set as the median of the Euclidean distances of all node pairs.

For ease of formulation and inspired by the work in [98], we define a scaled assignment vector $\mathbf{u}_v \in \mathbb{R}^T$ for each node of the tree over all the T leaf nodes which can be stated as

$$u_{vt} = \begin{cases} \frac{1}{\sqrt{|\mathcal{G}_v|}}, & \text{if } t \in \mathcal{G}_v, \\ 0, & \text{otherwise.} \end{cases} \tag{4.17}$$

Based on the scaled assignment $\mathbf{u}_v$ and the pairwise similarity matrix $\mathbf{S}$, we can further formulate the affinity e_v for the node v as follows:

$$e_v = \mathbf{u}_v^T \mathbf{S} \mathbf{u}_v. \tag{4.18}$$

Since the characteristics of the affinity definition, the value of the e_v is limited within the range of $[0, 1]$. More importantly, such affinity measurement can guarantee that higher nodes correspond to weaker relatedness, and vice versa.

4.4.3 COMPLEXITY ANALYSIS

In order to analyze the complexity of our proposed TRUMANN model, we have to estimate the time complexity for constructing $\mathbf{A}$, $\mathbf{B}$, and $\mathbf{W}$ as defined in Eqs. (4.5), (4.7), and (4.15). The computational complexity of the training process is $O(M \times (O_1 + O_2 + O_3))$, where O_1, O_2 and O_3, respectively, equal to $((D^s)^2 N + (D^s)^3 + (D^s)^2 K)S$, $(NK^2 + NDKS + K^3 + K^2T)$

and $(2NK^2 + K^3)T$. Thereinto, M is the iteration times of the alternative optimization, which is a small value less than 10 in our above analysis. N, T, S, K, and D, respectively, refer to the number of micro-videos, venue categories, modalities, latent dimension, and the total feature dimensions over all the modalities. Usually, we consider only a few number of modalities. S is hence very small. In our experimental settings, K and T are in the order of a few hundreds. Meanwhile, the number of feature dimension is about 5,000. Therefore, D^2 is greater than K^2T. In light of this, we can reduce the time complexity to be ND^2, which is faster than SVM, in terms of $O(N^3)$.

4.4.4   EXPERIMENTS

To valid the effectiveness of the first model TRUMANN, we conducted several experiments over a server equipped with Inter(R) Core(TM) CPU i7-4790 at 3.6 GHz on 32 Gb RAM, 8 cores and 64-bit Windows 10 operation system. To thoroughly measure our model and the baselines, we employed multiple metrics, namely macro-F1 and micro-F1 [55]. The averaging macro-F1 gives equal weight to each class-label in the averaging process, whereas the averaging micro-F1 gives equal weight to all instances in the averaging process. Both macro-F1 and micro-F1 metrics reach their best value at 1 and worst score at 0.

The experimental results reported in this paper were based on 10-fold cross-validation. In particular, the stratified cross-validation [130] was adopted to ensure all categories contain approximately the same percentage between training and testing samples. In each round of the 10-fold cross-validation, we split Dataset II into three chunks: 80% of the micro-videos (i.e., 194,505 videos) were used for training, 10% (i.e., 24,313 videos) were used for validation, and the rest (i.e., 24,313 videos) were held out for testing. The training set was used to adjust the parameters, while the validation set was used to avoid overfitting, i.e., verifying that any performance increase over the training dataset actually yields an accuracy increase over a dataset that has not been shown to the model before. The testing set was used only for testing the final solution to confirm the actual predictive power of our model with optimal parameters. Grid search was employed to select the optimal parameters with small but adaptive step size.

Performance Comparison among Models
We carried out experiments on Dataset II to compare the overall effectiveness of our proposed TRUMANN model with several state-of-the-art baselines.

- **SRMTL**: The Sparse Graph Regularization Multi-Task Learning method can capture the relationship between task pairs and further impose a sparse graph regularization scheme to enforce the related pairs close to each other [99].

- **regMVMT**: This semi-supervised inductive multi-view multi-task learning model considers information from multiple views and learns multiple related tasks simultaneously [190]. Besides, we also compared our model with the variant of **regMVMT** method, dubbed **reg-**

MVMT+. regMVMT+ can achieve better performance by modeling the non-uniformly related tasks.

- **MvDA+RMTL**: This baseline is the combination of Multi-view Discriminant Analysis [75] and Robust Multi-Task Learning [26]. In particular, **MvDA** seeks for a single discriminant common space for multiple views by jointly learning multiple view-specific linear transforms. Meanwhile, the **RMTL** is able to capture the task relationships using a low-rank structure via group-sparse lasso.

- **TRUMANN-**: This baseline is the variant of our proposed model by setting all e_v in Eq. (4.3) to be 1. In other words, this baseline does not incorporate the knowledge of the pre-defined hierarchical structure.

The comparative results are summarized in Table 4.1. From this table, we have the following observations: (1) TRUMANN achieves better performance, as compared to other multi-task learning approaches, such as SRMTL. This is because the SRMTL cannot capture the prior knowledge of task relatedness in terms of tree structure. On the other hand, it reflects that micro-videos are more separable in the learned common space; (2) multi-modal multi-task models, such as regMVMT and TRUMANN remarkably outperform pure multi-task learning models, such as SRMTL. This again demonstrates that the relatedness among multi-modalities can boost the learning performance; (3) the joint learning of multi-modal multi-task models, including regMVMT and TRUMANN, shows their superiors to the sequential learning of multi-view multi-task model, MvDA+RMTL. This tells us that multi-modal learning and multi-task learning can mutually reinforce each other; (4) we can see that TRUMANN outperforms TRUMANN-. This demonstrates the usefulness of the pre-defined hierarchical structure, and reveals the necessity of tree-guided multi-task learning; and (5) we conducted the analysis of variance (known as ANOVA) micro-F1. In particular, we performed paired t-test between our model and each of the competitors over 10-fold cross validation. We found that p-values are substantially smaller than 0.05, which shows that the improvements of our proposed model are statistically significant.

Representativeness of Modalities
We also studied the effectiveness of different modality combination. Table 4.2 shows the results. From this table, we observed that: (1) the visual modality is the most discriminant one among visual, textual, and acoustic modalities. This is because the visual modality contains more location-specific information than acoustic and textual modality. On the other hand, it signals that the CNN features are capable of capturing the prominent visual characteristics of venue categories; (2) the acoustic modality provide important cues for venue categories as compared to the textual modality across micro-F1 and macro-F1 metrics. But only given the acoustic modality, it is hard to estimate the venue categories for most of the videos, while the combination of visual and acoustic modality get an improvement than visual modality; (3) textual modality is

Table 4.1: Performance comparison between our model and the baselines on the venue category estimation over Dataset II (p-value*: p-value over micro-F1)

Models	Macro-F1	Micro-F1	P-value*
SRMTL	2.61 ± 0.19%	15.71 ± 0.21%	1.1e-3
regMVMT	4.33 ± 0.41%	17.16 ± 0.28%	7.0e-3
regMVMT+	4.53 ± 0.31%	18.35 ± 0.13%	9.1e-3
MvDA + RMTL	2.46 ± 0.18%	17.28 ± 1.67%	1.0e-3
TRUMANN-	3.75 ± 0.17%	24.01 ± 0.35%	1.0e-2
TRUMANN	**5.21 ± 0.29%**	**25.27 ± 0.17%**	–

Table 4.2: Representativeness of different modalities on Dataset II (p-value*: p-value over micro-F1)

Modality	Macro-F1	Micro-F1	P-value*
Visual	4.49 ± 0.09%	22.56 ± 0.10%	2.3e-2
Acoustic	2.79 ± 0.01%	16.25 ± 0.46%	2.9e-4
Textual	1.44 ± 0.29%	12.36 ± 0.38%	5.4e-4
Acoustic + Textual	2.87 ± 0.16%	16.86 ± 0.06%	6.4e-3
Visual + Acoustic	4.61 ± 0.08%	23.85 ± 0.20%	1.8e-2
Visual + Textual	4.52 ± 0.11%	23.54 ± 0.17%	1.1e-2
ALL	**5.21 ± 0.29%**	**25.27 ± 0.17%**	–

the least descriptive for venue category estimation. This is due to that the textual descriptions are noisy, missing, sparse, and even irrelevant to the venue categories; and (4) the more modalities we incorporate, the better performance we can achieve. This implies that the information of one modality is insufficient and multi-modalities are complementary to each other rather than mutually conflicting. This is a consensus to the old saying "two heads are better than one."

Case Studies

In Figure 4.2, we respectively list the top 8 categories with best performance in only visual modality, acoustic modality, textual modality, and their combination. From this figure, we have the following observations: (1) for visual modality, our model achieves stable and satisfactory performance on many venue categories, especially on those with discriminant visual characteristic, such as the micro-videos related to "Zoo" and "Beach;" (2) regarding the acoustic modality, our model performs better on those with regular sounds or noisy noise, such as "Music Venue"

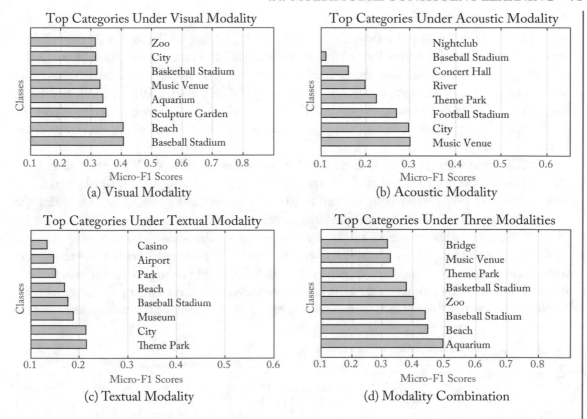

Figure 4.2: Categories with best classification performance under visual, acoustic, textual modality, and their combination, respectively. Experiments were conducted on Dataset II.

and "Concert Hall," which have discriminate acoustic signals as compared to other venue categories; (3) when it comes to the textual modality, we found that the top 8 best-performing categories are with high frequencies in micro-video descriptions. For instance, the terms of "Park" and "Beach" occur 2,992 and 3,882 times in our dataset, respectively. It is worth noting that not all the textual descriptions are correlated with the actual venue category, which in fact decreases the performance. For example, the textual description of one micro-video is *"I love my city."* Nevertheless, its venue category is "Park;" and (4) unsurprisingly, we obtained a significant improvement for "Aquarium" category, which is hard to recognize with only one modality. Moreover, compared to the performance over visual modality, the "Basketball Stadium" and "Zoo" categories are also improved about 8% in micro-F1. Besides, the more training samples one venue category contains, the higher probability of this category will yield, such as "Theme Park" and "City."

Parameter Tuning and Sensitivity

We have four key parameters as shown in Eq. (4.3): $K, \lambda_1, \lambda_2,$ and λ_3. The optimal values of these parameters were carefully tuned with 10-fold cross-validation in the training data. In particular, for each of the 10-fold, we chose the optimal parameters by grid search with a small but adaptive step size. Our parameters were searched in the range of [50, 500], [0.01,1], [0,1], and [0,1], respectively. The parameters corresponding to the best micro-F1 score were used to report the final results. For other competitors, the procedures to tune the parameters are analogous to the ensure fair comparison.

Take the parameter tuning in one of the 10-fold as an example. We observed that our model reached the optimal performance when $K = 200, \lambda_1 = 0.7, \lambda_2 = 0.4,$ and $\lambda_3 = 0.3$. We then investigated the sensitivity of our model to these parameters by varying one and fixing the others. Figure 4.3 illustrates the performance of our model with respect to $K, \lambda_1, \lambda_2,$ and λ_3. We can see that: (1) when fixing $\lambda_1, \lambda_2, \lambda_3$ and tuning K, the micro-F1 score value increases first and then reaches the peak value at $K = 200$; and (2) the micro-F1 score value changes in a small range, when varying $\lambda_1, \lambda_2,$ and λ_3 from 0–1. The slight change demonstrates that our model is non-sensitive to parameters. At last, we recorded the value of micro-F1 along with the iteration time using the optimal parameter settings. Figure 4.4 shows the convergence process

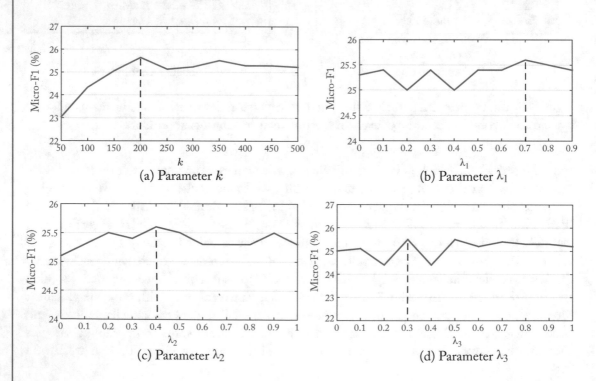

(a) Parameter k

(b) Parameter λ_1

(c) Parameter λ_2

(d) Parameter λ_3

Figure 4.3: Performance of TRUMANN on Dataset II with regards to varying parameters.

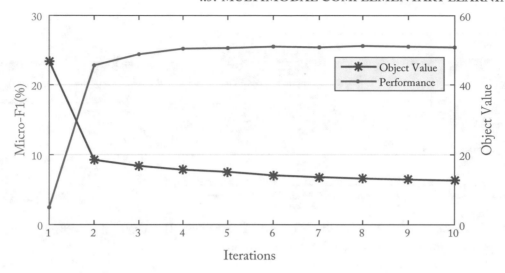

Figure 4.4: Performance of TRUMANN on Dataset II with the iteration times.

with respect to the number of iterations. From this figure, it can be seen that our algorithm can converge very fast.

4.5 MULTIMODAL COMPLEMENTARY LEARNING

It is worth mentioning that the pioneer work in Section 4.4 has studied the problem of micro-video categorization and devised a so-called "TRUMANN" model. However, TRUMANN explicitly projects all the modalities into the same feature space and represents all the modalities with a unified feature vector. In this way, the TRUMANN model does capture the common information among modalities, but it may lose some complementary information among them. For instance, the acoustic modality may contain the atom of "chirp of birds" that is hardly expressed by the visual modality. And the TRUMANN model aims to utilize the tree structure to guide the specific classifier learning rather than representation learning, it hence only suits for venue estimation task and cannot be applied to other applications. Moreover, the proposed TRUMANN model is an offline learning, which overlooks the importance of the online learning factor.

To address these problems, we develop an IncremeNtal Tree-guIded Multi-modAl dicTionary lEarning approach, dubbed INTIMATE, to organize micro-videos into a tree taxonomy. Our proposed approach is illustrated in Figure 4.5. Specifically, given a set of labeled micro-videos at the initial offline stage, our model is able to learn a concept-level dictionary for each modality, which is the basis of micro-video sparse representation. Standing on the shoulder

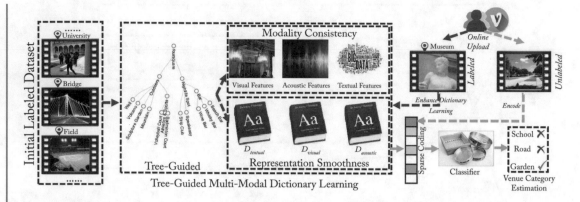

Figure 4.5: Scheme of our proposed INTIMATE approach. It consists of an offline dictionary learning component and an online learning component.

of the traditional dictionary learning framework, we advance it by devising a tree-guided group lasso via jointly considering the following two principles.

(1) Hierarchical Smoothness. Micro-videos with close labels in the hierarchical tree should have similar sparse representations.

(2) Structural Consistency. The tree structure is invariant across the textual, visual, and acoustic modalities. With the sparse representations, we can estimate the venue categories of micro-videos with shallow classifiers, such as softmax [30]. Moreover, we develop an online algorithm to solve the INTIMATE model. If an incoming micro-video is unlabeled, we can efficiently infer its venue category; otherwise, we will harvest its knowledge to strengthen our model.

In this part, we first briefly review the dictionary learning over mono-modal and multi-modal data. We then formulate the proposed INTIMATE model. At last, we optimize the INTIMATE model via an online algorithm.

4.5.1 MULTI-MODAL DICTIONARY LEARNING

Real-world objects are usually described by multi-modalities from different aspects. It is thus natural to extend the mono-modal dictionary learning to handle the multi-modal data. Suppose we have N samples with M modalities $\{(\mathbf{x}_n^1, \ldots, \mathbf{x}_n^M)\}_{n=1}^N$, in which $\mathbf{x}_n^m \in \mathbf{X}^m$ ($m = 1, \ldots, M$) denotes the m-th modality of the sample $\mathbf{x}_n$, $\mathbf{X}^m \in \mathbb{R}^{D_m \times N}$ denotes the m-th modality of the given N data, and D_m denotes the dimension of the m-th modality. The sparse representation of the n-th sample $\mathbf{A}_n = [\mathbf{a}_n^1, \ldots, \mathbf{a}_n^M] \in \mathbb{R}^{K \times M}$ and multi-modal dictionaries $\mathbf{D} = \{\mathbf{D}^1, \ldots, \mathbf{D}^M\}$

can be obtained by solving the reconstruction problem with $\ell_{2,1}$-norm,

$$\min_{\mathbf{A},\mathbf{D}} \frac{1}{2} \sum_{n=1}^{N} \sum_{m=1}^{M} \left\| \mathbf{x}_n^m - \mathbf{D}^m \mathbf{a}_n^m \right\|_2^2 + \lambda_1 \left\| \mathbf{A}_n \right\|_{2,1} + \frac{\lambda_2}{2} \left\| \mathbf{A}_n \right\|_F^2 ,$$
$$\text{s.t. } \mathbf{D}^m \in \mathbb{R}^{D_m \times K}, \ \left\| \mathbf{d}_j^m \right\| \leq 1, \ \forall j, m, \tag{4.19}$$

where $\mathbf{d}_j^m$ is the j-th column of $\mathbf{D}^m$, $\mathbf{a}_n^m$ is the sparse representation of the $\mathbf{x}_n^m$ over $\mathbf{D}^m$, and K is the number of atoms in each dictionary. Note that for a matrix $\mathbf{A}$, $\|\mathbf{A}\|_{2,1} = \sum_{i=1}^{m} \|\boldsymbol{\alpha}_i\|_2 = \sum_{i=1}^{m} \sqrt{\sum_{j=1}^{n} a_{ij}^2}$, where $\boldsymbol{\alpha}_i$ is the i-th row of $\mathbf{A}$ and a_{ij} is an element of $\mathbf{A}$ with the location at the i-th row and j-th column.

In the above equation, the $\ell_{2,1}$ group lasso is introduced to encourage row sparsity in $\mathbf{A}_n$, i.e., $\|\mathbf{A}_n\|_{2,1}$, it encourages collaboration among all the modalities by enforcing the same dictionary atoms from different modalities that present the same event, to reconstruct the input samples. The additional Frobenius norm $\| \cdot \|_F$ guarantees a unique solution to the joint sparse optimization problem.

4.5.2   TREE-GUIDED MULTI-MODAL DICTIONARY LEARNING

In our work, assuming we initially have N micro-videos for training. Each micro-video is described by M modalities and is exclusively associated with one of the T predefined venue categories (i.e., the leaf nodes of the tree. The internal nodes of the tree are much more abstract.). Our research objective is to learn a discriminant dictionary for each modality, denoted as $\mathbf{D}^m \in \mathbb{R}^{D_m \times K}$. Based upon these dictionaries, the sparse representation of the training samples are denoted as $\mathbf{A} = \{\mathbf{A}^1, \ldots, \mathbf{A}^M\}$, where $\mathbf{A}^m = [\mathbf{a}_1^m, \ldots, \mathbf{a}_N^m] \in \mathbb{R}^{K \times N}$ refers to the sparse representation of all the samples over the m-th modality.

As discussed above, the venue categories of micro-videos are organized into a tree $\mathcal{T}$ with a set of nodes $\mathcal{V}$, where the leaf nodes and the internal nodes, respectively, represent all the most specific venue categories and groups of the venue categories. Therefore, if we know the label of a given micro-video, we will know at which leaf node the video locates and hence its all ancestor nodes. Formally, each node $v \in \mathcal{V}$ of the tree has a group $\mathcal{G}_v = \{t_i\}$ consisting of all the leaf nodes t_i (venue categories), and it belongs to a subtree rooted at the node v. The micro-videos associated with the venue categories under the same node v tend to share a common set of concepts. Inspired by this, the venue categories under the same node are regularized to share a common set of dictionary atoms. It is worth emphasizing that the higher level node v locates, the fewer atoms can be shared. To characterize such property, we assign a weight e_v to each node $v \in \mathcal{V}$ according to the level it locates. It is noted that the root node possesses the highest level. We name such regularization as hierarchical smoothness. Besides, we ensure that different modalities share a common tree structure to guarantee the structural consistency.

Algorithm 4.2 Tree-Guided Multi-Modal Dictionary Learning Algorithm

---

Input:
    Initialization input matrix $\{\mathbf{X}^m\}_m^M$;
    Node assignment $\{\mathcal{G}_v\}_v^V$ with weights $\{e_v\}_v^V$;
    Parameters $\{K, \lambda, \gamma\}$;
Ensure:
    Discriminant dictionaries $\{\mathbf{D}^m\}_m^M$;
    Sparse coding $\{\mathbf{A}^m\}_m^M$;
  1: Initialize $t \leftarrow 0$;
  2: Initialize $\{\mathbf{A}_{(t)}^m\}_m^M$ and $\{\mathbf{D}_{(t)}^m\}_m^M$ randomly;
  3: **for** each modality m **do**
  4:     **while** $\mathbf{A}_{(t)}^m$ and $\mathbf{D}_{(t)}^m$ do not converge **do**
  5:         Fixing $\mathbf{A}_{(t)}^m$, construct each element of $\mathbf{q}_{(t)}^m$ using Eq. (4.24);
  6:         Fixing $\mathbf{D}_{(t)}^m$ and $\mathbf{q}_{(t)}^m$, update each column of $\mathbf{A}_{(t+1)}^m$ using Eq. (4.28);
  7:         Fixing $\mathbf{A}_{(t+1)}^m$, update $\mathbf{D}_{(t+1)}^m$ using Eq. (4.30);
  8:         update $t \leftarrow t + 1$;
  9:     **end while**
10:     **return** $\mathbf{D}^m \leftarrow \mathbf{D}_{(t)}^m$ and $\mathbf{A}^m \leftarrow \mathbf{A}_{(t)}^m$
11: **end for**

We formulate the multi-modal dictionary learning with a tree-constrained group lasso within a unified model Γ,

$$
\min_{\mathbf{D},\mathbf{A}} \frac{1}{2} \sum_{m=1}^M \left\| \mathbf{X}^m - \mathbf{D}^m \mathbf{A}^m \right\|_F^2 + \frac{\lambda}{2} \sum_{m=1}^M \sum_{v \in \mathcal{V}} e_v \left\| \mathbf{A}_{\mathcal{G}_v}^m \right\|_{2,1} + \frac{\gamma}{2} \sum_{m=1}^M \left\| \mathbf{A}^m \right\|_F^2 ,
$$
$$
\text{s.t. } \left\| \mathbf{d}_k^m \right\| \leq 1, \quad \forall k, m, \tag{4.20}
$$

where $\mathbf{A}_{\mathcal{G}_v}^m = \{\mathbf{a}_i^m : t_i \in \mathcal{G}_v\} \in \mathbb{R}^{K \times |\mathcal{G}_v|}$, $\mathbf{d}_k^m$ is the k-th column of $\mathbf{D}^m$, and the parameters e_v's are predefined. Suppose that each node v has n_v subnodes. The parameter e_v is heuristically set as n_v. We normalize $\{e_v\}_{v \in \mathcal{V}}$ by dividing them the maximum value of all the e_v's. With this normalization, we can map them into a range $[0,1]$.

Our objective function in Eq. (4.20) is composed of three terms: the first one is to measure the reconstruction error of each modality. The second term to force the micro-videos associated with the venue categories under the same node to share the similar sparse, i.e., hierarchical smoothness. We can see that $\mathcal{V}$ is invariant regarding m in the second term, which indeed implicitly ensures the structural consistency. The last term makes the objective function strongly convex and hence solvable.

Due to the fact that the three modalities are independent, we omit upper superscript m in the followings for better understanding. We rewrite the Eq. (4.20) as follows:

$$\min_{\mathbf{D},\mathbf{A}} \frac{1}{2}\|\mathbf{X} - \mathbf{D}\mathbf{A}\|_F^2 + \frac{\lambda}{2} \sum_{v \in \mathcal{V}} e_v \|\mathbf{A}_{\mathcal{G}_v}\|_{2,1} + \frac{\gamma}{2}\|\mathbf{A}\|_F^2, \quad \text{s.t. } \|\mathbf{d}_k\| \leq 1, \quad \forall k. \tag{4.21}$$

4.5.3 OPTIMIZATION

We adopt the alternating optimization strategy to solve the two variables in the Eq. (4.21) until convergence.

Computing $\mathbf{A}$ with $\mathbf{D}$ fixed: Considering that the second term in Eq. (4.21) is not differentiable, we thus turn to its equivalent formulation, which has been proven by the work of [3], $\frac{\lambda}{2}(\sum_{v \in \mathcal{V}} e_v \|\mathbf{A}_{\mathcal{G}_v}\|)^2$. It is, however, still intractable. We further turn to another variational formulation of this equation. According to the Cauchy-Schwarz inequality, given an arbitrary vector $\mathbf{b} = [b_1,\ldots,b_L]^T \in \mathbb{R}^L$ such that $\mathbf{b} \neq \mathbf{0}$, we have

$$\sum_{i=1}^{L} |b_i| = \sum_{i=1}^{L} \theta_i^{\frac{1}{2}} \theta_i^{-\frac{1}{2}} |b_i| \leq \left(\sum_{i=1}^{M} \theta_i^{-1} b_i^2 \right)^{\frac{1}{2}}, \tag{4.22}$$

where θ_i's are the introduced variables that satisfy $\sum_{i=1}^{L} \theta_i = 1, \theta_i > 0$ and the equality holds when $\theta_i = |b_i|/\|\mathbf{b}\|_1$.

Based on this lemma, we derive the following inequality:

$$\left(\sum_{v \in \mathcal{V}} e_v \|\mathbf{A}_{\mathcal{G}_v}\| \right)^2 \leq \sum_{k=1}^{K} \sum_{v \in \mathcal{V}} \frac{e_v^2 \|\mathbf{A}_{k,\mathcal{G}_v}\|_2^2}{q_{k,v}}, \tag{4.23}$$

where $\sum_k \sum_v q_{k,v} = 1, q_{k,v} \geq 0, \forall k, v$ and $\mathbf{A}_{k,\mathcal{G}_v}$ denotes the k-th row of the matrix $\mathbf{A}_{\mathcal{G}_v}$. The equality holds when

$$q_{k,v} = \frac{e_v \|\mathbf{A}_{k,\mathcal{G}_v}\|_2}{\sum_{k=1}^{K} \sum_{v \in \mathcal{V}} e_v \|\mathbf{A}_{k,\mathcal{G}_v}\|_2}. \tag{4.24}$$

Thus far, we have theoretically derived that minimizing Γ with respect to $\mathbf{A}$ is equivalent to minimizing the following convex object function:

$$\min_{\mathbf{A},q_{k,v}} \frac{1}{2}\|\mathbf{X} - \mathbf{D}\mathbf{A}\|_F^2 + \frac{\gamma}{2}\|\mathbf{A}\|_F^2 + \frac{\lambda}{2} \sum_{k=1}^{K} \sum_{v \in \mathcal{V}} \frac{\|e_v\mathbf{A}_{k,\mathcal{G}_v}\|_2^2}{q_{k,v}}. \tag{4.25}$$

To facilitate the computation of the derivation of the objective function Γ regarding $\mathbf{a}_n$ for the n-th training example, we define a diagonal matrix $\mathbf{Q}^n \in \mathbb{R}^{K \times K}$ for the training sample n over the corresponding modality. Its diagonal entry is defined as

$$\mathbf{Q}_{kk}^n = \sum_{\{v \in \mathcal{V}|y_n \in v\}} \frac{e_v^2}{q_{k,v}}, \tag{4.26}$$

where y_n denotes the category label of the sample $\mathbf{x}_n$. We ultimately reach the following objective function:

$$\min_{\mathbf{A},\mathbf{Q}} \frac{1}{2} \sum_{n=1}^{N} \|\mathbf{x}_n - \mathbf{D}\mathbf{a}_n\|_F^2 + \frac{\gamma}{2} \sum_{n=1}^{N} \|\mathbf{a}_n\|_F^2 + \frac{\lambda}{2} \sum_{n=1}^{N} (\mathbf{a}_n)^T \mathbf{Q}^n \mathbf{a}_n. \tag{4.27}$$

The alternative optimization strategy is applicable here. By fixing $\mathbf{Q}^n$, taking derivative of the above formulation regarding $\mathbf{a}_n$, and setting it to zero, we reach

$$\begin{cases} -\mathbf{D}^T (\mathbf{x}_n - \mathbf{D}\mathbf{a}_n) + \gamma \mathbf{a}_n + \lambda \mathbf{Q}^n \mathbf{a}_n = 0, \\ \left(\mathbf{D}^T\mathbf{D} + \gamma \mathbf{I} + \lambda \mathbf{Q}^n\right) \mathbf{a}_n = \mathbf{D}^T \mathbf{x}_n, \\ \mathbf{a}_n = \left(\mathbf{D}^T\mathbf{D} + \gamma \mathbf{I} + \lambda \mathbf{Q}^n\right)^{-1} \left(\mathbf{D}^T \mathbf{x}_n\right). \end{cases} \tag{4.28}$$

Once we obtain all the $\mathbf{a}_n$, we can easily compute $\mathbf{Q}^n$ based on Eqs. (4.24) and (4.26).

Computing $\mathbf{D}$ with $\mathbf{A}$ fixed: Fixing $\mathbf{A}$ and taking the derivative of Γ with respect to $\mathbf{D}$, we have

$$\frac{\partial \Gamma}{\partial \mathbf{D}} = (\mathbf{D}\mathbf{A} - \mathbf{X})\mathbf{A}^T. \tag{4.29}$$

By setting Eq. (4.29) to zero, it can be derived that

$$\begin{cases} \mathbf{D}\left(\mathbf{A}\mathbf{A}^T\right) - \left(\mathbf{X}\mathbf{A}^T\right) = 0, \\ \mathbf{D} = \left(\mathbf{X}\mathbf{A}^T\right)\left(\mathbf{A}\mathbf{A}^T\right)^{-1}. \end{cases} \tag{4.30}$$

It is straightforward that the above algorithm converges, because in each iteration, Γ will decrease, as shown in Figure 4.6. By using Algorithm 4.2, we can learn a set of dictionaries $\mathbf{D}^m$ for each modality of samples $\mathbf{X}^m$ and their corresponding representations $\mathbf{A}^m$.

4.5.4   ONLINE LEARNING

As analyzed in the introduction, the efficient operation and incremental learning of micro-videos deserve our attention. To accomplish this, we present an online learning algorithm (referred to Algorithm 4.3). Generally speaking, if an incoming sample is labeled, we leverage it to strengthen the dictionary learning. We treat the learned $\mathbf{D}$ over the initial training data as $\mathbf{D}_{(0)}$ and update it to $\mathbf{D}_{(t)}$ at the current time t. Otherwise, we compute its sparse representation based on the current dictionaries and classify it into the right venue category.

An Incoming Labeled Sample: At the t-th online update, a new sample $\mathbf{x}_t$ with a label $\mathbf{y}_t$ is given. We can know which leaf node this micro-video is from and then use it to update the dictionaries $\mathbf{D}_{(t-1)}$. From Eq. (4.30), we find that the solution of $\mathbf{D}_{(t)}$ relies on the sparse representation $\mathbf{A}_{(t)} = [\mathbf{A}_{(t-1)}, \mathbf{a}_t]$. We thus need to compute $\mathbf{a}_t$ first that is the representation vector of $\mathbf{x}_t$. However, Eq. (4.28) tells us that $\mathbf{a}_t$ is related to $\mathbf{Q}^t$ computed by $\mathbf{A}_{(t)} = [\mathbf{A}_{(t-1)}, \mathbf{a}_t]$.

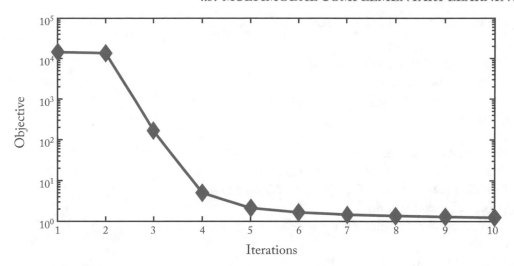

Figure 4.6: Example of the convergence of Algorithm 4.2.

To address this problem, we firstly initialize $\mathbf{a}_t$ to get a temporal $\mathbf{A}_{(t)} = [\mathbf{A}_{(t-1)}, \mathbf{a}_t]$, and then we use Eq. (4.26) to compute $\mathbf{Q}^t$. Afterward, we can use Eq. (4.28) to compute $\mathbf{a}_t$ with $\mathbf{D}_{(t-1)}$ as the dictionary. We repeat this procedure until we obtain the stable $\mathbf{A}_{(t)}$ for sample $\mathbf{x}_t$.

To estimate $\mathbf{D}_{(t)}$ when fixing $\mathbf{A}_{(t)}$, we adopt the similar procedure introduced in [109]. In particular, we sequentially update each column of $\mathbf{D}_{(t)}$. We here take the j-th column as an example to illustrate the procedure.

We define $\mathbf{d}_j(t)$ as the j-th column of $\mathbf{D}_{(t)}$. And we set

$$g\left(\mathbf{D}_{(t)}\right) = \frac{1}{2} \sum_{i=1}^{t} \left\| \mathbf{x}_i - \mathbf{D}_{(t-1)}\mathbf{a}_i \right\|_2^2 . \tag{4.31}$$

We then set $\nabla_{\mathbf{d}_j(t)} g(\mathbf{D}_{(t)})$ to be zero, and obtain

$$\mathbf{d}_j(t) = \frac{\sum_{i=1}^{t} a_{ij}^T \left(\mathbf{x}_i - \tilde{\mathbf{D}}\tilde{\mathbf{a}}_i \right)}{\sum_{i=1}^{t} a_{ij}^T a_{ij}}, \tag{4.32}$$

where a_{ij} is the j-th entry of $\mathbf{a}_i$, $\tilde{\mathbf{D}} = \mathbf{D}_{(t-1)} \setminus \{\mathbf{d}_j(t\text{-}1)\}$ is a dictionary excluding the j-th atom and $\tilde{\mathbf{a}}_i = \mathbf{a}_i \setminus \{a_{ij}\}$ defines the coefficients for the corresponding dictionary atoms of $\tilde{\mathbf{D}}$.

After deriving this equation, we have the additive property of linear solution

$$\frac{\sum_{i=1}^{t} a_{ij}^T \left(\mathbf{x}_i - \mathbf{D}_{(t-1)}\mathbf{a}_i \right)}{\sum_{i=1}^{t} a_{ij}^T a_{ij}} + \mathbf{d}_j(t-1). \tag{4.33}$$

Algorithm 4.3 Our INTIMATE Algorithm

Input:
 Initialization input matrix $\{\mathbf{X}^m\}_m^M$;
 Streaming data $\{\ldots, \mathbf{x}_t, \ldots\}$;
 Node assignment $\{\mathcal{G}_v\}_v^V$ with weights $\{e_v\}_v^V$;
 Parameters $\{\lambda, \gamma\}$;

Ensure:
 Discriminant dictionaries $\{\mathbf{D}^m\}_m^M$;
 Sparse coding $\{\mathbf{a}_t^m\}_m^M$ of $\mathbf{x}_t$ and its label;
1: Initialize $\{\mathbf{D}_{(0)}^m\}_m^M$ and $\{\mathbf{A}_{(0)}^m\}_m^M$ using Algorithm 4.2;
2: **for** each modality m **do**
3:   Training the classifier f^m using $\mathbf{A}_{(0)}^m$;
4: **end for**
5: Initialize $t \leftarrow 1$;
6: **for** a newly sample $\mathbf{x}_{(t)}$ in the stream **do**
7:   **if** $\mathbf{x}_{(t)}$ has a label $\mathbf{y}_{(t)}$ **then**
8:    **for** each modality m **do**
9:     Fixing $\mathbf{D}_{(t-1)}^m$, learn $\mathbf{a}_{(t)}^m$ using Eq. (4.28);
10:     Fixing $\mathbf{A}_{(t-1)}^m$ and $\mathbf{a}_{(t)}^m$, update $\mathbf{D}_{(t)}^m$ using Eq. (4.35) and Eq. (4.36);
11:    **end for**
12:   **else if** $\mathbf{x}_{(t)}$ without label **then**
13:    **for** each modality m **do**
14:     Learning the representation $\mathbf{a}_{(t)}^m$ with $\mathbf{D}_{(t-1)}^m$;
15:     Leveraging $\mathbf{a}_{(t)}^m$ and f^m, predict its label $\mathbf{y}_t^m$;
16:    **end for**
17:   Based on $\{\mathbf{y}_t^m\}$, obtain the final label $\mathbf{y}_t$ using Eq. (4.37);
18:   **end if**
19:   update $t \leftarrow t + 1$;
20: **end for**
21: **return** $\mathbf{D}^m \leftarrow \mathbf{D}_{(t)}^m$

We set

$$\begin{cases} \mathbf{U}(t) & = [\mathbf{u}_1(t), \ldots, \mathbf{u}_K(t)] = \mathbf{U}(t-1) + \mathbf{a}_t \, (\mathbf{a}_t)^T, \\ \mathbf{F}(t) & = [\mathbf{f}_1(t), \ldots, \mathbf{f}_K(t)] = \mathbf{F}(t-1) + \mathbf{x}_t \, (\mathbf{a}_t)^T, \\ \mathbf{U}(0) & = \sum_{n=1}^N \mathbf{a}_n(0) \, (\mathbf{a}_n(0))^T, \\ \mathbf{F}(0) & = \sum_{n=1}^N \mathbf{x}_n(0) \, (\mathbf{a}_n(0))^T, \end{cases} \quad (4.34)$$

where $\mathbf{a}_n(0)$ is the n-th column of $\mathbf{A}_{(0)}$, and $\mathbf{x}_n(0)$ is the n-th column of the offline sample matrix $\mathbf{X}$.

We then can rewrite Eq. (4.33) as

$$\mathbf{d}_j(t) = \frac{1}{\mathbf{U}_{jj}(t)}\left(\mathbf{f}_j(t) - \mathbf{D}_{(t-1)}\mathbf{u}_j(t)\right) + \mathbf{d}_j(t-1). \tag{4.35}$$

Because $\|\mathbf{d}_j(t)\| \leq 1$, we need to use

$$\mathbf{d}_j(t) = \frac{1}{\max\left(\|\mathbf{d}_j(t)\|, 1\right)}\mathbf{d}_j(t), \tag{4.36}$$

to normalize the atom $\mathbf{d}_j(t)$.

An Incoming Unlabeled Sample: If the new sample $\mathbf{x}_t$ does not have a label, we have to predict its label. We first utilize the previous learned dictionaries to obtain the sparse representations for the sample $\mathbf{x}_t$. In the offline stage, we utilize the sparse representations of training samples to learn a Softmax model for venue category classification. We use the trained classifiers to predict the label of the new sample $\mathbf{x}_t$. Specifically, we judge the label of the sample with the following method:

$$\mathbf{y} = \arg\min_{\mathbf{q}^t, t\in\{1,\dots,T\}} \sum_{m=1}^{M} \left\|\mathbf{q}^t - \mathbf{y}^m\right\|_2^2, \tag{4.37}$$

where $\mathbf{q}^t$ is a binary vector for the t-th category, whereby its t-th coordinate is one and the remaining ones are zeros, and $\mathbf{y}^m$ is the predicted label based on the m-th modality.

Due to the fact that we adopt the optimization method proposed in [109] and the convergence of the algorithm has been proved in their paper, here we omit the proof of the convergence.

4.5.5   EXPERIMENTS

Since our proposed model is an online dictionary learning algorithm, we do not need too much data to train the offline model. In order to demonstrate the robustness of our model, we only selected a small set of data from the Dataset II as our training data. Specifically, we randomly selected 5,396 micro-videos as our offline training data to learn the dictionaries, 10,807 micro-videos as our online training data, and 2,170 micro-videos as our testing data. We repeated the random selection ten times and reported the average experimental results.

Baselines

To shed light on the effectiveness of our proposed approach, we compared it with the following several state-of-the-art baselines.

- Without dictionary learning (**WDL**): We trained the softmax classifier directly with the raw features without considering the dictionary learning.

- Sparse graph regularization multi-task learning (**SRMTL**): This is a multi-task learning model capturing the structural relatedness between task pairs [99]. This model is also trained with raw features.

- Data-driven dictionary learning (**DDL**): This is a classic mono-modal unsupervised dictionary learning framework utilizing elastic-net [35].

- Online dictionary learning for sparse coding (**ODLSC**): This is an online mono-modal dictionary learning relying on the stochastic approximations. It is capable of scaling up to millions of training samples [109].

- Multi-modal unsupervised dictionary learning (**MDL**): It is a joint sparse representation model via $\ell_{2,1}$ norm (as formulated in Eq. (4.19)).

- Multi-modal task-driven dictionary learning (**MTDL**): This is a multi-modal task-driven dictionary learning algorithm [6] enabling modalities to collaborate at both the feature level by using joint sparse representation and the decision level by using a sum of the decision scores.

- Tree-guided multi-task multi-modal learning (**TRUMANN**): It learns a common space from the multi-modal data and utilizes it to represent each micro-video. Then, it leverages a multi-task learning model to predict the category of a micro-video [192].

Overall Performance Comparison

We trained our approach and the baselines over the offline training set and verified them over the testing one. The results are comparatively summarized in Table 4.3. We observed that: (1) DDL

Table 4.3: Efficient and effective performance comparison between our model and the baselines. We also reported the average results over 10-round experiments based pairwise significance test. (p-value*: p-value over accuracy)

Models	Accuracy	Micro-F1	Time(s)	P-value*
WDL	2.79 ± 0.07%	2.89 ± 0.04%	65.0	$3.05e\text{-}16$
SRMTL	1.79 ± 0.05%	1.47 ± 0.10%	140.7	$8.40e\text{-}19$
DDL	2.95 ± 0.09%	3.12 ± 0.10%	1,194.7	$1.51e\text{-}15$
ODLSC	2.92 ± 0.15%	3.04 ± 0.17%	1,046.2	$1.97e\text{-}14$
MDL	4.43 ± 0.06%	4.66 ± 0.07%	4,468.9	$7.19e\text{-}12$
MTDL	4.50 ± 0.13%	4.75 ± 0.15%	3,338.6	$4.50e\text{-}10$
TRUMANN	4.19 ± 0.02%	4.46 ± 0.03%	**51.4**	$9.14e\text{-}14$
INTIMATE	**6.28 ± 0.08%**	**6.60 ± 0.09%**	150.1	–

and ODLSC methods outperformed WDL and SRMTL. The latter two methods leveraged the raw features to directly train the classifier, whereas the former two learned the sparse representations of micro-videos before training the classifiers. This justifies the necessity of dictionary learning and the discrimination of sparse representation; (2) multi-modal dictionary learning methods surpass the mono-modal ones. This demonstrates that there are indeed relatedness among modalities. Appropriate capturing and modeling such relatedness can reinforce the dictionary learning, and hence the discrimination of sparse representation; (3) multi-modal dictionary learning methods achieve relatively better results than that of TRUMMAN. This signals that not all the modalities of micro-videos share the same space and it thus may not be the best to learn the common space via the agreement constrains; (4) MTDL shows its superiority to the MDL. This tells us that in the supervised settings, joint minimization of misclassification and reconstruction errors result in the dictionaries that are adapted to the desired tasks. And supervised methods can lead to a more accurate classification compared with the unsupervised ones; (5) our proposed INTIMATE substantially outperforms the others, including MTDL and MDL. This verifies that harvesting the prior knowledge of tree structure is useful for a more discriminative dictionary learning toward venue classification; and (6) as to the efficiency over the offline part, we can see that the cost of our model is relatively lower than that of the others. In addition, we also conducted the significance test between our model and each of the baselines relying on the average results over 10-round experiments cross validation results. We can see that all the p-values are substantially smaller than 0.05, indicating that the advantage of our model is statistically significant. Besides, we can see that the performance of our model hardly fluctuates, showing the stability of our model.

Parameter Tuning and Sensitivity Analysis
Our model has three key parameters: the number of dictionary atoms K, the tradeoff parameters λ and γ. In each of the 10-round experiments, we adopted the grid search strategy to carefully tune and select the optimal parameters from the training data [123]. Take one experiment as an example. We first performed the grid search in a coarse level within a wide range of [0, 1000] using an adaptive step size. Once we obtained the approximate scope of each parameter, we then performed the fine tuning within a narrow range using a small step size.

Figure 4.7 shows the performance of our model regarding the three parameters, which is accomplished by varying one and fixing the others. We can see that the performance of our model changes within small ranges nearby the optimal settings. This justifies that our model is non-sensitive to the parameters around their optimal settings. It is observed that the setting of $K = 150$, $\lambda = 0.85$, and $\gamma = 1$ works well for all of our experiments. The procedures of tuning the parameters are analogous to other competitors to ensure a fair comparison.

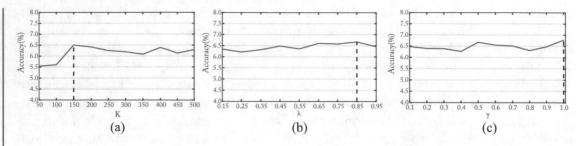

(a)   (b)   (c)

Figure 4.7: Parameter tuning and sensitivity analysis. This is implemented by varying one parameter and fixing the rest.

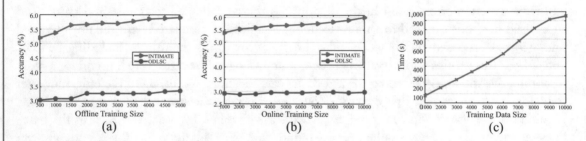

(a)   (b)   (c)

Figure 4.8: (a) Influence of offline training size on the online update results. (b) Influence of online training size on the accuracy of online update stage. (c) Time consuming of offline stage with different sizes of the training data.

Justification of Online Learning

In this experiment, we have three settings: (1) by fixing the offline training data (5,396 videos), we studied the effectiveness of the online part (increasing the samples from 1–10,807); (2) fixing the online training data (10,807 videos), we studied the size influence of the offline training data (increasing the samples from 1–5,396); and (3) we studied the time consuming of the offline stage with different sizes of the training data, and compared the time with that of the online stage.

The comparison results over 2,170 testing samples between INTIMATE and ODLSC are illustrated in Figures 4.8a and 4.8b. We have the following observations: (1) the more online training samples used to train our model, the better performance it achieves. This justifies the usefulness of leveraging the incoming samples to incrementally strengthen our model. Meanwhile, it stably outperforms ODLSC under the same experimental setting; and (2) the size of the initial training samples does affect the performance of our model, but not significantly even if the online samples are insufficient. This shows the robustness of our model.

Figure 4.8c shows the time consuming of the offline stage with different sizes of the training data. From this figure, we find that the amount of time increases monotonously with the size of the training data. Namely, if a new labeled sample is given (now the training data is 10,001), it may consume at least 900 s to retrain the offline part for testing. However, our online learning model only needs less than 1 s to strengthen our model for testing. It certifies that the online learning is very necessary.

Effectiveness of Parallel Dictionary Learning

In order to evaluate the effectiveness of our parallel multi-modal dictionary learning, we proposed a share dictionary learning model. The work in [44] has justified that a deep convolutional neural network is equivalent to the sparse dictionary learning pipeline, whereby each convolutional filter can be seen as a dictionary atom that we aim to learn, and the sparse coding can be seen as the activation value of the filtered results. Motivated by this work, we designed a deep share dictionary learning model as shown in Figure 4.9. Due to the fact that all the modality features should utilize the same dictionary, we first embedded the three modality features into a common space. Suppose the dimension of the common space is n, we then constructed a dictionary using $n \times 1$ CNN kernels. Representation learning over a dictionary with K atoms is equivalent to applying K linear filters $n \times 1$ to each input feature vector (each column in the matrix). The sparse coding solver will then iteratively process K coefficients.

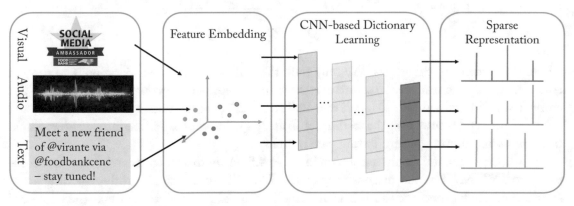

Figure 4.9: The pipeline of the baseline model DPL. It firstly embeds the three modality features into a common space, and then uses a CNN-based dictionary learning method to learn sparse representations. Finally, it concatenates these representations and throws them into a classifier.

The experimental results are shown in Table 4.4. From this table, we find that: (1) DPL performs better than the above baseline methods, including MTDL and MDL. This demonstrates the superior performance of deep learning-based model. (2) Although this model is a deep learning one, our model still outperforms it. Because this model projects all the modalities into a common space for sharing a common dictionary, it loses the complementary information among the modalities. This also verifies that the common space assumption proposed in [192] is invalid.

Table 4.4: Comparison between models with CNN-based dictionary learning and our dictionary learning for the venue category estimation (p-value*: p-value over accuracy)

Models	Accuracy	Micro-F1	P-value*
DPL	4.64 ± 0.24%	4.87 ± 0.28%	3.86e-08
INTIMATE	**6.28 ± 0.08%**	**6.60 ± 0.09%**	–

Effectiveness of the Tree Structure

We argued that encoding the tree structure to constrain the sparse representation learning can strengthen the representation discrimination. In this part, we carried out experiments to verify the effectiveness of the tree structure from quantitative and qualitative aspects.

Quantitative Analysis: To show the effect of the tree structure on the sparse representation learning, we compared it with a flat model without tree structure, dubbed INTIMATE$^-$,

$$\min_{\mathbf{D},\mathbf{A}} \frac{1}{2} \sum_{m=1}^{M} \|\mathbf{X}^m - \mathbf{D}^m \mathbf{A}^m\|_F^2 + \frac{\lambda}{2} \sum_{m=1}^{M} \sum_{c \in \mathcal{C}} \|\mathbf{A}_c^m\|_{2,1} + \frac{\gamma}{2} \sum_{m=1}^{M} \|\mathbf{A}^m\|_F^2 , \qquad (4.38)$$
$$\text{s.t. } \|\mathbf{d}_k^m\| \leq 1, \quad \forall k, m,$$

where $\mathcal{C}$ is the set of categories. We only took the leaf nodes into consideration, and did not consider the hierarchical tree structure to regularize the representation learning. To ensure a fair comparison, we trained INTIMATE and INTIMATE$^-$ over the same offline training set and reported the final results over the testing set. Analogous to other experiments, we also repeated this one on ten round training/testing data sampling.

The experimental results are shown in Table 4.5. From this table, it can be seen that compared to INTIMATE, the performance of INTIMATE$^-$ drops significantly regarding accuracy and micro-F1 metrics. This is because, the INTIMATE$^-$ baseline does take the class label information into consideration and learns the category-aware sparse representations for each micro-video, however, it completely ignores the hierarchical relatedness among categories. This further justifies the usefulness of encoding the tree structure to learn the sparse representations.

Table 4.5: Comparison between models with and without structure information for the venue category estimation on Dataset II (p-value*: p-value over accuracy)

Models	Accuracy	Micro-F1	P-value*
INTIMATE$^-$	3.00 ± 0.04%	3.18 ± 0.05%	6.86e-15
INTIMATE	**6.28 ± 0.08%**	**6.60 ± 0.09%**	–

Qualitative Analysis: Apart from the quantitative analysis, we also conducted the qualitative one to intuitively show the effects of the hierarchical tree. As illustrated in Figure 4.10, we selected an internal node (i.e., "Outdoors & Recreation") and descendants from the given tree. It is worth noting that to save the display space, we deliberately selected the internal node close to the leaf ones. For each leaf node under the selected "Outdoors & Recreation" node, we randomly sampled a few testing micro-videos categorized by our model for demonstration.

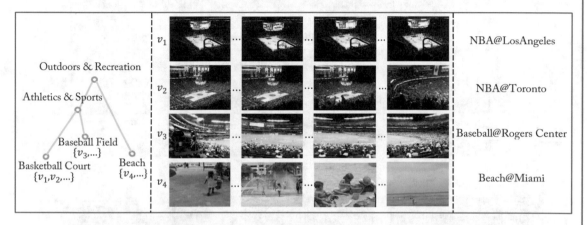

Figure 4.10: Qualitative analysis of the structure effectiveness. We can see that visually similar micro-videos have geographically close categories.

In Figure 4.10, in comparison to v_1 and v_3, we can see that v_1 and v_2 are in the same leaf node "Basketball Court," and they share most of the visual concepts and are the most visually similar pairs. Meanwhile, as compared to v_1 and v_4, we can see that v_1 and v_3 are visually closer. Therefore, we can conclude that the geographically closer the venue categories of two micro-videos are, the more visually similar they will be. This observation strongly supports our assumption of hierarchical smoothness.

Justification of Modality Combination

We also studied the performance of our model with different modality combinations. The results are summarized in Table 4.6. It can be seen that: (1) the visual modality outperforms the acoustic and textual ones. This is due to the fact that the visual modality is more intuitive to show venue information than that of the acoustic and textual ones; (2) combining the visual modality with acoustic modality outperforms the combination of visual modality and the textual modality. This reflects that the acoustic modality conveys more important cues on venue categories than the textual one does. Because the textual descriptions are of low-quality, noisy, incomplete, sparse, and even irrelevant to the venue categories; (3) the single modality is insufficient to estimate the venue category, but combining them can largely enhance the performance; and (4) our proposed

INTIMATE achieves the best performance over three modalities. This further justifies that multi-modalities are complementary instead of conflicting.

Table 4.6: Performance of our proposed INTIMATE model with different modality combinations on Dataset II (p-value*: p-value over accuracy)

Modality	Accuracy	Micro-F1	P-value*
Visual	5.02 ± 0.14%	5.30 ± 0.18%	$1.31e\text{-}07$
Audio	4.78 ± 0.13%	5.08 ± 0.17%	$8.96e\text{-}09$
Text	4.61 ± 0.12%	4.96 ± 0.14%	$8.68e\text{-}10$
Visual + Audio	5.46 ± 0.18%	5.65 ± 0.18%	$8.92e\text{-}05$
Visual + Text	5.15 ± 0.24%	5.54 ± 0.18%	$7.47e\text{-}06$
Audio + Text	5.05 ± 0.14%	5.48 ± 0.24%	$1.90e\text{-}07$
All	**6.28 ± 0.08%**	**6.60 ± 0.09%**	–

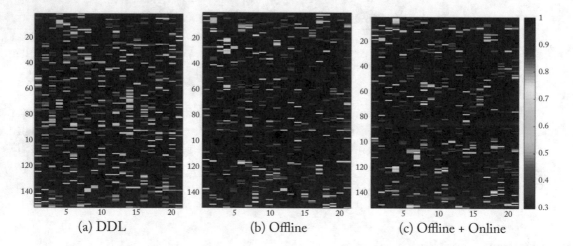

(a) DDL (b) Offline (c) Offline + Online

Figure 4.11: Visualizing the sparse representations of some micro-videos. The y-axis denotes the atom IDs and the x-axis stands for the micro-video exemplar IDs. Brighter bars refer to higher weights.

Visualization of Sparse Representation

We visualized the representations of some examples. In particular, we first randomly selected two internal nodes that are geometrically far away in the tree. Following that, we randomly selected 10 micro-videos from each of the two nodes from the testing set. We then visualized their sparse representations of the visual modality as shown in Figure 4.11. Video ID 1-10 in the x-axis are from one venue category and 11-20 are from another. We can see that: (1) sparse representations achieved by DDL in Figure 4.11a are somehow independent and not very sparse; (2) the sparse representations based on our INTIMATE model trained on the offline training data are more sparse as shown Figure 4.11b; and (3) after enhanced by the online data, we can see from Figure 4.11c that samples from the same node often sparsely share the same set of atoms.

Categorization Examples

To gain the deep insights into our proposed INTIMATE model, we illustrated two categorization results of micro-videos in Figure 4.12.

The top micro-video in Figure 4.12 contains band and colorful lights. And we heard sound of music and singing from its acoustic channel. Obviously, the venue category of this one is

Text Description: #mattyhealy #fillmoredetroit #the 1975 #detroit.

Ground Truth: Concert Hall   Predict Venue: Concert Hall

Text Description: kayaking!

Ground Truth: Garden   Predict Venue: Garden

Figure 4.12: Exemplars of micro-videos categorization.

"Concert Hall." Although the textual modality contains useless information, our model can utilize the information from the visual and acoustic modalities to guide the estimation of the venue category. Similarly, from the bottom micro-video in Figure 4.12, it can be seen that many people kayaking on a lake surrounding by a lot of trees. Therefore, it was captured at "Garden." Our proposed model successfully predicts the venues of these micro-videos, which demonstrates that our model can learn discriminative representations to distinguish well micro-videos of different venue categories.

4.6   MULTIMODAL COOPERATIVE LEARNING

Technically speaking, venue category estimation of micro-videos is usually treated as a multimodal fusion problem, and solved by integrating the geographic cues from visual, acoustic, and textual modalities of micro-videos. Several pioneering efforts have been dedicated to the task, such as [102, 192], and [122]. The current methods, however, are restricted to only fusing the common (a.k.a., consistent) cues over multiple modalities or complementary cues. Moving one step forward, in this work, we shed light on the cooperative relations, comprising the consistent and complementary components. We refer to the consistent component as the same information appearing in more than one modality in different forms. As shown in Figure 4.13, a red candy displaying in the visual modality and the text of "lollipop" describe the consistency. By contrast, the complementary component represents the exclusive information appearing only in one modality. For instance, it is hard to find the equivalent in other modalities in Figure 4.13 of the textual concept of "girl" or the visual concept of "grass." To supercharge a multimodal prediction scheme with such cooperative relations, the multimodal cooperation shall be able to: (1) enhance the confidence of the same evidence from various views via consistent regularization;

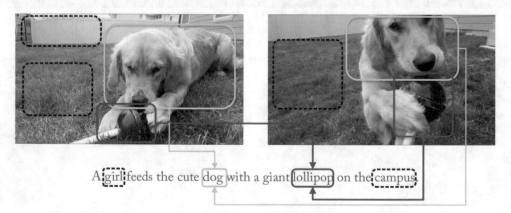

A girl feeds the cute dog with a giant lollipop on the campus

Figure 4.13: Exemplar demonstration of the correlation between the visual modality and textural modality. The blue and brown boxes show the consistent information and the red dashed boxes show the complementary ones, respectively.

and (2) provide a comprehensive representation from the exclusive perspective of complementary component. Nevertheless, characterizing and modeling multimodal cooperation are non-trivial due to the following challenges: (1) consistent and complementary information is often mixed. How to separate it from different modalities is largely untapped; and (2) after separation, it is difficult to associate them with each other, since they are orthogonal.

To address the problems analyzed above, we present an end-to-end deep multimodal cooperative learning approach to estimating the venue categories of micro-videos. Notably, this approach is applicable to other multimodal cooperative scenarios. As illustrated in Figure 4.14, the features are firstly extracted from each modality and fed into three peer cooperative nets. In

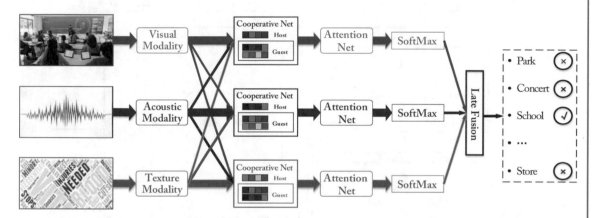

Figure 4.14: An illustration of our framework. It separates the consistent features from the complementary ones and enhances the expressiveness of each modality via the proposed cooperative net. Then, it selects the features to generate a discriminative representation in the attention network toward venue category estimation.

each cooperative net, we respectively treat one modality as the host and the rest as the guests. Then we obtain the augmented feature vectors as the output of the cooperative nets. Following that, each vector is fed into an attention net followed by a late fusion over the prediction results from different softmax functions. Stepping into the cooperative net as demonstrated in Figure 4.15, the structure is symmetric. In particular, on the left hand, we first concatenate the guest modalities and estimate the relevance between each dimension of the combining vector and the host vector. As to the combined vector, a gate with the learned threshold is used to separate its consistent part and complementary part. An analogous process is applied to the right hand side. Thereafter, two consistent parts are fused with a deep neural network model and the fusion result is ultimately concatenated with the two complementary parts.

We first formally define the problem. Assume that we are given a set of N micro-videos $\mathcal{X} = \{x_i\}_{i=1}^N$. For each micro-video $x \in \mathcal{X}$, we segment it into three modalities $\{\mathbf{x}_v, \mathbf{x}_a, \mathbf{x}_t\}$, where v, a, and t denote the visual, acoustic, and textual modality indices, respectively. Let

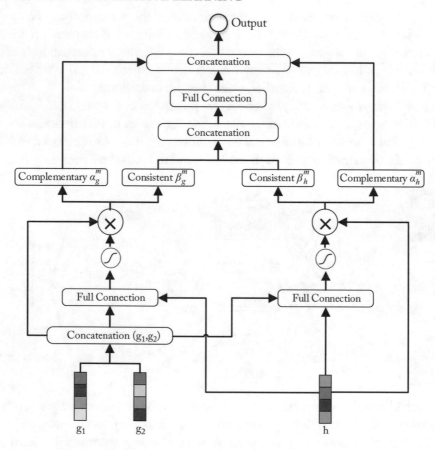

Figure 4.15: Illustration of Cooperative Net. The cooperative nets separete the consistent components from the complmentary ones, and yield an augmented feature vector comprised of the enhanced consistent vector and complementary vectors.

$m \in \mathcal{M} = \{v, a, t\}$ denote the modality indicator, and $\mathbf{x}^m \in \mathbb{R}^{D_m}$ denote the D_m-dimensional feature vector over the m-th modality. In our work, each micro-video is associated with one of K pre-defined venue categories, namely a one-hot label vector $\mathbf{y} \in \mathbb{R}^K$, where K refers to the number of venue category.

4.6.1 MULTIMODAL EARLY FUSION

As aforementioned, fusing multimodal information is capable of producing the comprehensive description for micro-videos. Prior studies [47, 166] have practically demonstrated the effectiveness of early fusion strategy, which concatenates the features from all modalities into a unified representation. Following that, one can devise a classifier, such as a neural network, treating the overall representation as the input. We term the generic solution as *multimodal early fusion*. Formally, for each micro-video x, we concatenate $\mathbf{x}^v$, $\mathbf{x}^a$, and $\mathbf{x}^t$ into one vector as

$$\mathbf{x} = \left[\mathbf{x}^v, \mathbf{x}^a, \mathbf{x}^t\right], \tag{4.39}$$

where $\mathbf{x}$ is the multimodal representation by fusing the features from the visual, acoustic, and textual modalities.

In fact, the early fusion implicitly assumes that the modalities are linearly independent, overlooking the correlations among the modalities. Hence, it fails to explore the modal correlations to strengthen the expressiveness of each modality and further improve the capacity of the fusion method. In this work, we argue that the information across modalities can be categorized into two parts: consistent and complementary components. For example, let certain features of $\mathbf{x}^v$ indicate the visual concepts of "sunshine" and "crowed;" and some of $\mathbf{x}^a$ describe the acoustic concepts of "wind" and "crowed cheering." From the angle of consistency, the visual concept of "crowed" is consistent with the acoustic concept of "crowed cheering." For complementarity, the visual concept of "sunshine" provides the exclusive signals, as compared to the acoustic one of "wind."

Uncovering the underlying modality relations in micro-videos is already challenging, not to mention different types of relations to the final prediction. To the best of our knowledge, most existing efforts only implicitly model the modality relations during the learning process, leaving the explicit exhibition of relations untouched. Specifically, the deep learning based methods, which feed multimodal features together into a black box multi-layer neural network and output a joint representation, are widely used to characterize the multimodal data. With the deep neural network, the correlations between different features are involved in the new representations. However, the corresponding features cannot be captured and filtered from the vectors. Toward this end, we aim to propose a novel cooperative learning mechanism to leverage the uncovered relations and boost the prediction performance.

4.6.2 COOPERATIVE NETWORKS

Our preliminary consideration is to explicitly model the relations comprised of the consistent and complementary parts. A viable solution [1, 133] is to project the representations of different modalities into a common latent space. In this solution, the consistent cues should be close to each other since they show the same evidences, whereas the complementary cues in the common space should be distant due to the fact they have no overlapping information. To map the heterogeneous information extracted from a micro-video to the same coordinate, some information, especially the modality specific information, probably lose during the projection. We term it as the common-specific method. Hence, such direct mapping will lead to suboptimal expressiveness of the method. Although through careful parameter tuning, we can control the loss to a certain extent, it requires extensive experiments which are not easily adapted to other applications.

To avoid such information loss, we devise a novel solution named cooperative network, in which each modality information is overall retained and augmented by the other modalities. Specifically, this network assigns each dimension of features with a relation score and consequently divides the features into consistent and complementary parts. Here the relation score for each feature reflects how consistent the information is derived from the other modalities. The use of relation score endows our model with strong expressiveness and benefits the further cooperative learning. In what follows, we elaborate the key ingredients of cooperative network.

Relation Score

The goal of the relation score is to select features from each modality, where the underlying information is consistent among modalities. As shown in Figure 4.15, we treat one specific modality m as the host represented as $\mathbf{h}^m$; the other modalities as the guests denoted as $\mathbf{g}_1^m$ and $\mathbf{g}_2^m$, respectively. Intuitively, we can explicitly capture the varying consistency of the host and guest features by assigning an attentive weight for each feature dimension. The weights are considered as the relation scores. Therefore, given the representations of the host and guest modalities, we present a novel relation-aware attention mechanism to score each feature.

Considering that the consistency should be the correlation between the host and the whole guest modality, we concatenate all the guest vectors together as follows:

$$\mathbf{g}^m = \left[\mathbf{g}_1^m, \mathbf{g}_2^m\right], \tag{4.40}$$

where the $\mathbf{g}^m$ encapsulates all the features from the guest modalities. Subsequently, we feed the guest vector $\mathbf{g}^m$ and the host vector $\mathbf{h}^m$ into the attention scoring function, which is a neural network composed of a single hidden layer and a softmax layer. And the output of this function is a host score vector, where the value of each dimension reflects the degree of a host feature derived from the whole guest features. The degree reaches the highest at 1 and the lowest at 0. It is formally defined as

$$\mathbf{s}_h^m = \text{softmax}\left(\mathbf{W}_h^m \cdot [\mathbf{h}^m, \mathbf{g}^m]\right), \tag{4.41}$$

where $\mathbf{W}_h^m \in \mathbb{R}^{D_h * D}$ and $\mathbf{s}_h^m \in \mathbb{R}^{D_h}$ denote the weight matrix and relation score vector corresponding to each dimension of the host vector, respectively; the D_h denotes the dimension of the host vector, and D is the dimension of the overall vector. For simplicity, we omit the bias terms.

For the guest modality, we analogously score the feature dimensions to measure the degree of a guest feature derived from the host features, defined as follows:

$$\mathbf{s}_g^m = \text{softmax}\left(\mathbf{W}_g^m \cdot [\mathbf{g}^m, \mathbf{h}^m]\right), \tag{4.42}$$

where $\mathbf{W}_g^m \in \mathbb{R}^{D_g \times D}$, the D_g and $\mathbf{s}_g^m \in R^{\mathbb{R}_g}$ denote the weight matrix, the dimension of guest vector and the relation score vector corresponding to each dimension of the guest vector, respectively.

Consistency and Complementary Components

Having established the attentive relation scores, we can easily locate the consistent and complementary features from each modality. Toward this end, we set a trainable threshold denoted as ξ_o^m, in which we use $o \in \mathbb{O} = \{h, g\}$ as the host and guest indicator. This threshold divides the relation score vector into two parts: consistent vector and complementary vector, namely γ_o^m and δ_o^m. The element in the complementary vectors is defined as follows:

$$\delta_o^m[i] = \begin{cases} 1 - s_o^m[i], & \text{if } s_o^m[i] < \xi_0^m, \\ 0, & \text{otherwise,} \end{cases} \tag{4.43}$$

where $\delta_o^m[i]$ is the value of the i-th dimension in the complementary weight vector δ_o^m, reflecting the degree of the complementary relation. For the consistent weight vector γ_o^m, we formulate its element as

$$\gamma_o^m[i] = \begin{cases} s_o^m[i], & \text{if } s_o^m[i] \geq \xi_0^m, \\ 0, & \text{otherwise,} \end{cases} \tag{4.44}$$

where $\gamma_o^m[i]$ is the value of i-th the dimension in the consistent weight vector γ_o^m, indicating the degree of the consistency.

Particular, since the original functions are not continuous, we introduced a sigmoid function to make them differentiable, as follow:

$$\begin{cases} \gamma_o^m[i] = \dfrac{s_o^m[i]}{1 + e^{-w*(s_o^m[i] - \xi_o^m)}}, \\ \delta_o^m[i] = 1 - \gamma_o^m[i], \end{cases} \tag{4.45}$$

where w denotes a scalar weighting the difference between $s_o^m[i]$ and ξ_o^m to make the output $\gamma_o^m[i]$ as close as possible to 0 or $s_o^m[i]$. Through experiments, the best results are obtained with a weight of 50.

After that, we gain four correlation weight vectors from each host-guest pair, namely δ_h^m, δ_g^m, γ_h^m, and γ_g^m. Based on these weight vectors, we separated the consistent features and the complementary features from the mixed information, which are the element-wise products of the original feature vector and each weight vector as,

$$\begin{cases} \alpha_h^m = \mathbf{h}^m \otimes \delta_h^m, \\ \alpha_g^m = \mathbf{g}^m \otimes \delta_g^m, \\ \beta_h^m = \mathbf{h}^m \otimes \gamma_h^m, \\ \beta_g^m = \mathbf{g}^m \otimes \gamma_g^m, \end{cases} \tag{4.46}$$

where two complementary vectors and two consistent vectors of the host modality and guest modality are denoted as α_h^m, α_g^m, β_h^m, and β_g^m, respectively.

With the separated consistent and complementary components, we can reconstruct the representations with better expressiveness. We employ different strategies on distinct components. To adequately exploit the correlations between the consistent component pairs, we concatenate these vectors and feed them into a neural network to learn an enhanced consistent vector,

$$\tilde{\beta}^m = \varphi\left(\mathbf{W}_{\beta}^m \cdot [\beta_h^m, \beta_g^m]\right), \tag{4.47}$$

where $\mathbf{W}_{\beta}^m$, $\varphi(\cdot)$, and $\tilde{\beta}^m$ denote the weight matrix, activation function, and the enhanced consistent vector in the modality m, respectively.

To supplement the exclusive information from other modalities, we integrate the enhanced consistent components and the complementary components to generate a feature vector with powerful expressiveness as

$$\hat{x}^m = \left[\alpha_h^m, \ \tilde{\beta}^m, \ \alpha_g^m\right]. \tag{4.48}$$

Meanwhile, to guarantee the consistency, the diversity of the consistent component pairs should be minimized. However, the dimension of each vector is different, and the number of the consistent features is dynamic. We hence fail to capture the diversity of these features directly. Toward this end, we propose to compute the probability distributions of venue categories represented by consistent vectors, and further leverage the Kullback–Leibler divergence (KL divergence) to encourage them to be close.

Particularly, the probability distribution over categories is defined as follows:

$$\mathbf{p}_o^m = \text{softmax}\left(\mathbf{U}_o^m \cdot \beta_o^m\right), \tag{4.49}$$

where $\mathbf{U}_o^m \in \mathbb{R}^{K*D_o}$ and $\mathbf{p}_o^m \in \mathbb{R}^K$, respectively, denote the weight matrix and the probability distribution of the venue categories represented by the consistent vector β_o^m.

Following that, we compute the KL divergence between the two probability distributions $\mathbf{p}_h^m$ and $\mathbf{p}_g^m$, formally as,

$$\mathcal{L}_1^m = \sum_{x \in \mathcal{X}} (\mathbf{p}_g^m \log \mathbf{p}_h^m - \mathbf{p}_h^m \log \mathbf{p}_g^m), \tag{4.50}$$

where $\mathbf{p}_h^m$ and $\mathbf{p}_g^m$ both denote the probability distribution of the venue categories. Based upon this, we calculate the sum of the KL divergences from all modalities as

$$\mathcal{L}_1 = \sum_{m \in \mathcal{M}} \mathcal{L}_1^m. \tag{4.51}$$

4.6.3　ATTENTION NETWORKS

Given the augmented representations above, a straightforward way to estimate the venue category is to adopt a classifier. However, we argue that the rich information within the augmented representations is redundant for the prediction task, and hence the simple classifier can hardly select the discriminative features. Several efforts have been paid to achieve the a discriminative representation from massive features, like PCA [159] and sparse representation [207]. These approaches, however, have many hyper-parameters to tune. More principle components, for instance, can lead to the suboptimal performance.

With the advance of the attention mechanism, we employ an attention network to evaluate the attention scores for each feature toward different venue categories. These scores can measure the relevance and significance of the features for the venue category. In addition, the continuous attention scores make the feature selection flexible. Thereafter, we obtain the scored features and leverage them to learn a discriminative representation to estimate the venue category in each modality.

Attention Score

Given a feature vector, we assign an attention score to each feature according to the venue category and obtain the scored feature to learn a discriminative representation.

Instead of computing the importance of each feature to categories, we constructed a trainable memory matrix to store the attention score of them. The matrix is denoted as $\mathbf{\Omega}^m \in \mathbb{R}^{D_m \times K}$ in the modality m and the entry in row i and column j represents the importance of i-th feature toward j-th venue category . For each category, the scored feature vector is obtained by calculating the element-wise product of the feature vector and the corresponding row vector in matrix $\mathbf{\Omega}^m$. It is formulated as

$$\boldsymbol{\psi}_j^m = \boldsymbol{\omega}_j^m \otimes \hat{\mathbf{x}}^m, \tag{4.52}$$

where $\hat{\mathbf{x}}^m \in \mathbb{R}^D$ is the augmented vector in the modality m; $\boldsymbol{\omega}_j^m \in \mathbb{R}^D$ denotes the feature attention scores of the venue category j and $\boldsymbol{\psi}_j^m \in \mathbb{R}^D$ denotes the scored feature vector toward venue category j.

To yield the discriminative representation, we feed the scored feature vector into a fully connected layer as follows:

$$\boldsymbol{\theta}_j^m = \phi(\mathbf{W}^m \cdot \boldsymbol{\psi}_j^m), \tag{4.53}$$

where $\mathbf{W}^m$, $\phi(\cdot)$, and $\boldsymbol{\theta}_j^m$ denote the weight matrix, the activation function and the discriminative representation of j-th venue category in the modality m, respectively.

Multimodal Estimation

After obtaining the discriminative representations, we pass them into a fully connected softmax layer. It computes the probability distributions over the venue category labels in each modality, mathematically stated as

$$p\left(\hat{y}_k^m \mid {}_k^m\right) = \frac{\exp\left(\mathbf{z}_k^{\mathrm{T}} \boldsymbol{\theta}_k^m\right)}{\sum_{k'=1}^{K} \exp\left(\mathbf{z}_{k'}^{\mathrm{T}} \boldsymbol{\theta}_{k'}^m\right)}, \tag{4.54}$$

where $\mathbf{z}_k$ is a weight vector of the k-th venue category, and $\boldsymbol{\theta}_k^m$ can be viewed as the discriminative representation of k-th venue category in the modality m. Thereafter, we obtain the probabilistic label vector $\hat{\mathbf{y}}^m = [\hat{y}_1^m, \cdots, \hat{y}_K^m]$ over K venue categories.

For multiple modalities, we fuse the probabilistic label vector over three modalities, defined as follows:

$$\hat{\mathbf{y}} = \sum_{m \in \mathcal{M}} (\hat{\mathbf{y}}^m). \tag{4.55}$$

Following that, we adopt a function to minimize the loss between the estimated label vector and its target values, as

$$\mathcal{L}_2 = -\sum_{x \in \mathcal{X}} \sum_{k=1}^{K} \mathbf{y}_k \log(\hat{\mathbf{y}}_k). \tag{4.56}$$

Ultimately, this function and the KL divergence of consistent representation pairs are combined as the objective function of our proposed method, as follows:

$$\mathcal{L} = \mathcal{L}_1 + \mathcal{L}_2. \tag{4.57}$$

4.6.4 EXPERIMENTS

We validated our proposed NMCL model and its components over micro-video understanding.

Experiment Settings

In addition to Macro-F1 and Micro-F1, we provide the Receiver Operating Curves (ROC) of our method and four baselines and use Areas Under Curve (AUC) scores to evaluate the results. AUC is the area under the ROC curve, which is created by plotting the true positive rate against the false positive rate. We divided our Dataset II into three chunks: 132,370 for training, 56,731 for validation, and 81,044 for testing. The training set is used to adjust the parameters, while the validation one provides an unbiased evaluation of a model fit on the training dataset and tunes the model's hyperparameters. The testing one is used only to report the final solution to confirm the actual predictive power of our model with the optimal parameter settings.

We compared the performance of our proposed model with several state-of-the-art baselines.

- **Early Fusion** [119]: For any given micro-video, we concatenated multimodal features into one vector, and then learned a model consisting of three fully connected layers to estimate the venue category over the concatenation vectors.

- **Late Fusion** [119]: To calculate the categories distribution, we devised the classifiers which are respectively implemented by a neural network with one, two, and three hidden layers for the textual, acoustic, and visual modalities. And we fused these distributions to yield a final prediction venue category.

- **Early+Att** [64]: This baseline is the combination of the early fusion and attention model. In particular, the attention model gives different attention weights to all features integrated from multiple modalities according to different venue categories. Here, the attention weights are calculated by a scoring function of the concatenated features and venue category. After that, a neural network is devised with three fully connected layers to categorize the unseen micro-videos over the attended feature vectors.

- **Late+Att** [64]: For various venue categories, features in each modality have varying contributions to the final prediction. Therefore, this baseline introduces the attention mechanism into classifiers of each modality to obtain the venue category representations and then fuses these representations to yield a final venue category.

- **TRUMANN**: This is a tree-guided multi-task multi-modal learning method introduced in Section 4.4, which is the first one toward the micro-video venue category estimation. This model is able to jointly learn a common space from multiple modalities and leverage the predefined Foursquare hierarchical structure to regularize the relatedness among venue categories.

- **DARE** [122]: This work is a deep transfer model which harnesses the external knowledge to enhance the acoustic modality and regularizes the representation learning of micro-videos of the same venue category to alleviate the sparsity problem of unpopular categories.

We implemented our model with the help of Tensorflow.[1] Particularly, we applied the Xavier approach to initialize the model parameters, which has been proved as an excellent initialization method for the neural network models. The mini-batch size and learning rate are respectively searched in {128, 256, 512} and {0.001, 0.005, 0.01, 0.05, 0.1}. The optimizer is set as Adaptive Moment Estimation (Adam) [80]. Moreover, we empirically set the size of each hidden layer as 256 and the activation function as ReLU. Without special mention, all the models employ one hidden layer and one prediction layer. For a fair comparison, we initialized other competitors with an analogous procedure. The average results over five-round predictions are illustrated in the testing set.

Performance Comparison

The comparative results are shown in Table 4.7 and Figure 4.16. From this table, we have the following observations:

1. In terms of the Micro-F1, Early Fusion and Late Fusion achieve the worst performance, since these standard fusion approaches rarely exploit the correlations between different modalities.

---

[1]https://www.tensorflow.org

Table 4.7: Performance comparison between our model and the baselines (p-value*: p-value[2] over micro-F1)

	Micro-F1	Macro-F1	P-value*
Early Fusion	11.39 ± 0.01%	0.12 ± 0.01%	1.31e-8
Late Fusion	12.57 ± 0.23%	0.20 ± 0.04%	4.29e-9
Early + Att	31.24 ± 0.37%	14.03 ± 0.19%	2.48e-8
Late + Att	30.00 ± 0.31%	13.71 ± 0.51%	1.52e-7
TRUMANN	27.38 ± 0.21%	10.87 ± 0.05%	8.71e-8
DARE	34.40 ± 0.32%	20.21 ± 0.35%	5.94e-7
NMCL	**40.04 ± 0.37%**	**26.78 ± 0.42%**	–

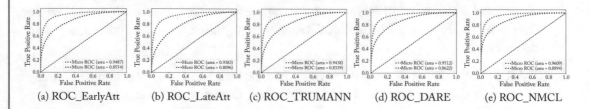

(a) ROC_EarlyAtt   (b) ROC_LateAtt   (c) ROC_TRUMANN   (d) ROC_DARE   (e) ROC_NMCL

Figure 4.16: ROC curves and AUC scores of methods.

2. Integrating the attention model to the standard fusion is able to improve the performance obviously. Taking the advantages of the attention mechanism, Early+Att and Late+Att can dynamically select the discriminative features, which are tailored to the prediction task. This verifies the feasibility of revising the weight of each feature.

3. When performing the estimation task, TRUMANN outperforms Early Fusion and Late Fusion. It is reasonable since it considers the hierarchical structure of venue categories and employs the multi-task learning, whereas Early+Att and Late+Att outperform the TRUMANN. It again admits the effectiveness of assigning the attentive weights to the features.

4. The performance of DARE exceeds the others except ours, indicating that DARE benefits from the enhanced audio modality via an external dataset and alleviates the sparse problem of unpopular categories by regularizing the similarity among the categories.

5. Our proposed model achieves the best w.r.t. micro-F1 and macro-F1. By exhibiting the consistency and complementary of features, our model achieves a better expressiveness

---

[2]In statistical hypothesis testing, the probability value (p-value) is the probability for a given statistical model that, when the null hypothesis is true, the statistical summary would be greater than or equal to the actual observed results.

compared to all baselines. While DARE and TRUMANN treat all features linearly, independently, and equally, our model can capture and leverage the correlation between different modalities, as well as employ the attention networks to identify the tailored attention of each feature. We further conducted a pair wise significant test to verify that all improvements are statistically significant with p-value < 0.05.

6. As shown in Figure 4.16, NMCL archives an AUC score of more than 96% and is superior to the baselines, further demonstrating the effectiveness of our proposed method. Despite DARE yields the AUC score of 95.12% and ranks the second-best performance among all the methods, our proposed method outperforms it by a gain of about 1%. Besides, in terms of the macro-average ROC curve, NMCL gets an AUC score of about 89%, which increases by 3–10% than the baselines.

Study of NMCL Model

We studied the effectiveness of combining different modalities. Tables 4.8 and 4.9 show the performance of different modality pairs and each enhanced modality with our proposed model, respectively. In addition, we plotted the Macro-F1 and Micro-F1 *w.r.t.* the number of iterations in Figure 4.17 to illustrate the convergence and efficiency of our model in each modality.

From these tables and the figure, we observed the following.

1. On the first row in Table 4.8, solely considering the visual modality achieves the best performance compared to the other mono-modal estimation methods. This is consistent with the finding in [122, 192], verifying the rich geographic information conveyed by the visual features. In addition, the CNN features are capable of capturing the prominent visual characteristics of the venue categories.

2. The acoustic modality and textual modality perform similarly in estimating the venue categories, which are listed on the second row and the third row in Table 4.8, respectively. Only

Table 4.8: Representativeness of different modalities

	Micro-F1	Macro-F1
Textual	13.40 ± 0.14%	2.23 ± 0.1%
Acoustic	14.21 ± 0.12%	3.40 ± 0.02%
Visual	28.16 ± 0.23%	11.22 ± 0.41%
Acoustic + Textual	20.57 ± 0.41%	7.08 ± 0.09%
Visual + Textual	38.45 ± 0.34%	23.83 ± 0.34%
Visual + Acoustic	37.07 ± 0.35%	23.34 ± 0.11%
All	**40.04 ± 0.37%**	**26.78 ± 0.42%**

Table 4.9: Performance of each enhanced modality in different modality pairs. (V-MicroF1, A-MicroF1, and T-MicroF1 denote Micro-F1 score on the visual, acoustic, and textual modality, respectively.)

	V-Micro-F1	A-Micro-F1	T-Micro-F1
Acoustic + Textual	–	20.12 ± 0.15%	20.13 ± 0.14%
Visual + Textual	**37.46 ± 0.26%**	–	**37.75 ± 0.36%**
Visual + Acoustic	35.09 ± 0.15%	34.8 ± 0.16%	–
All	36.07 ± 0.28%	**35.27 ± 0.17%**	33.73 ± 0.51%

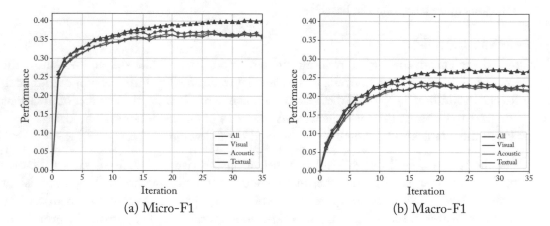

(a) Micro-F1   (b) Macro-F1

Figure 4.17: Convergence and effective study of the NMCL.

using one modality, however, is insufficient to estimate the categories for most micro-videos, since the textual and acoustic information is noisy, sparse, and even irrelevant to the venue categories.

3. The more modalities we incorporate, the better performance we can achieve, as the display on last three rows in Tables 4.8 and 4.9. This implies that the information of one modality is insufficient and multiple modalities are complementary to each other rather than conflicting. This is a consensus to the old saying "two heads are better than one."

4. Table 4.9 shows that the performance of each modality enhanced by our proposed approach is improved obviously, especially the acoustic and textual modalities are combined with the visual modality. This improvement validates that each modality can be enforced by the other modalities in our model.

5. Comparing each row in Table 4.9 to the first three rows in Table 4.8, the performance of each modality, which is enhanced by the other two modalities with our model, is better than that of the early fusion integrating the attention model. It indicates that our model can capture the correlation between different modalities.

6. Jointly analyzing the curves in Figure 4.17, we found that utilizing our proposed cooperative learning to seamlessly integrate multiple modalities can boost the performance effectively. This demonstrates the rationality of our model. And the performance tends to be stable at around 30 iterations. This signals the convergence property of our model and also indicates its efficiency.

We also list several variants based on our proposed cooperative net. These methods group the features of each modality into consistent and complementary parts. Afterwards, we adopted different fusing strategies to leverage the consistent and complementary features, including the following.

- **Variant-I**: In this model, Eq. (4.47) is removed. In other words, we integrated the guest complementary information into the host modality without enhancing the consistent parts, while the guest consistent part is retained to calculate the KL-diversity for keeping the consistency.

- **Variant-II**: This variant discards Eq. (4.48) and merely harnesses the consistent vector pairs to learn an enhanced feature vector for each modality and categorize the venue with these enhanced feature vectors.

- **Variant-III**: After obtaining the consistent and complementary features from each host and modality pair, Eq. (4.47) is replaced, and a new enhanced consistent vector is learned by integrating all host consistent vectors. After that, the category is estimated by fusing the predictions of the newly enhanced consistent vector and each complementary vector.

- **Variant-IV**: In this variant model, we respectively concatenated all complementary parts and all consistent parts, instead of Eqs. (4.47) and (4.48). Finally, we estimated the venue category of the two concatenated parts and fused them to gain the result.

From Table 4.10, we have the following observations.

1. In terms of Macro-F1, Variant-I and Variant-IV outperform Variant-II and Variant-III, respectively. This may because combining the complementary information can involve more information, strengthening the expressiveness of the representations.

2. The accuracy of the first two variants is comparatively higher than the other variants. This benefits from capturing the correlation between the host and the guest features which is ignored by the Variant-III and the Variant-IV.

Table 4.10: Performance of variants

	Micro-F1	Macro-F1
Variant-I	39.17 ± 0.27%	25.05 ± 0.28%
Variant-II	39.01 ± 0.37%	23.70 ± 0.19%
Variant-III	38.11 ± 0.40%	22.78 ± 0.10%
Variant-IV	38.48 ± 0.33%	24.49 ± 0.18%
NMCL	**40.04 ± 0.37%**	**26.78 ± 0.42%**

3. Our proposed method outperforms its all variants, justifying the rationality and effective-ness of cooperative learning. Different from several variants, the original one considers the consistency between each host and guest modality pairs and supplements the exclusive signals from the guest modalities.

4. We observe that Variant-I, which discards one of the consistent parts, does not cause a significant reduction in accuracy. It shows that the information contained in the two con-sistent vectors is almost the same, and it also proves that our model can correctly distinguish and capture the consistent features.

5. Comparing the proposed method with Variant-II, we observe that the improvement in terms of Micro-F1 is not significant. For further analysis, we believe that the main rea-son is that the concepts contained in micro-videos are sparse. Moreover, the information contained in any single modality is almost covered by the other two modalities. In other words, the complementary parts contain little external information. Therefore, the removal of the complementary parts barely affects the performance.

Visualization

Apart from achieving more accurate prediction, the key advantage of NMCL over other meth-ods is that it exhibits the consistent and complementary features. Toward this end, we show examples drawn from our model to visualize two representation components.

Since the acoustic modality is the hardest one to be visualized among the multiple modal-ities, we utilized the concept-level features to present the acoustic one. To extract the concept from fine-grained acoustic features, we leveraged an external dataset namely AudioSet, which is a large-scale dataset released by Google.[3]

The AudioSet consists of an expanding ontology of 632 audio event classes and a collection of 2,084,320 human labeled 10-s sound clips drawn from YouTube videos. The ontology is specified as a hierarchical graph of event categories, covering a wide range of human and animal

[3]The external audio dataset was just used for the visualization. https://research.google.com/audioset/.

sounds, musical instruments and genres, and everyday common sounds from the environment, like "Speech," "Laughter," and "Guitar."

To estimate the concepts in the audio, we employed a VGG-like model [62] and trained it over the AudioSet. According to the input format of the CNN model, we regenerated the acoustic features of the micro-videos. The extracted audios are divided into non-overlapping 960 ms frames, and then the spectrogram transformed from the frames are integrated into 64 mel-spaced frequency bins. Finally, we took the mean pooling strategy over all the frames of the micro-video to yield a new acoustic feature vector.

With the new acoustic conceptual features, we conducted experiments to shed some light on the correlation between the acoustic modality and the other modalities. In addition, we visualized the attention score matrix between the acoustic concepts and venue categories to validate our proposed model intuitively.

- To visualize the consistent and the complementary components, we selected exemplary demonstrations of two micro-videos categorized as "Park" and "Piazza place," as shown in Figure 4.18a and 4.18b. For these demonstrations, we treated the acoustic modality as the host part, the visual and the textual modalities as the guests. And we showed a heat map to illustrate the correlation between the host and guest feature pairs, where the darker color indicates that the host feature is consistent with the guest modalities and vice versa. From Figure 4.18a, we observe that several acoustic concepts are consistent with the visual and textual modalities, such as "Music" and "Violin," and some are exclusive ones hardly revealed from the other modalities, such as "Applause," "Noise," and "Car alarm." In contrast, given Figure 4.18b, we find that the correlation score distribution is totally different. The concepts, such as "Applause," "Crowed," and "Noisy," can be represented by the guest

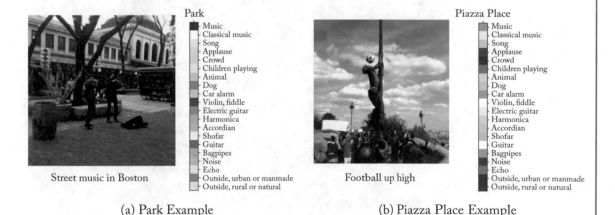

(a) Park Example   (b) Piazza Place Example

Figure 4.18: Visualization of the correlation scores between the same acoustic concept-level features and different visual and textual features.

features, and the "Music" and the "Violin" are barely captured in the other modalities. However, these "lighter-colored" features provide the exclusive and discriminative information to predict the venue category. In our proposed model, we can explicitly capture the exclusive information as a supplement, rather than omitting it during the learning produce. These observations verify the assumption that the information from different modalities is complementary to each other and demonstrate that our proposed model can explicitly separate the consistent information from the complementary one.

- To save the space, we performed the part of the attention matrix via a heat map, where the lighter color indicates weak attention and vice versa, as shown in Figure 4.19. We can see that every selected venue category has various relations to each acoustic concept. For instance, the micro-videos with the venue of "Mall" have strong correlations with "Speech" and "Children shouting;" the correlation with "Babbling" is loose. In addition, for the

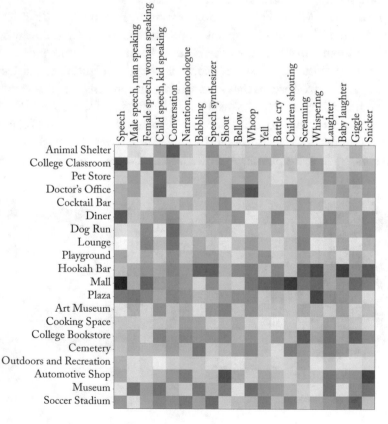

Figure 4.19: Visualization of the attention scores of acoustic concept-level features and venue category pairs.

venue of "Pet store," the colors representing "Kid speaking" and "Whoop" are dark, and the color representing "Battle cry" is lighter. These observations agree with our common sense and demonstrate that the attention score can select the discriminative features toward the venue category.

4.7   SUMMARY

This chapter presents three novel and efficient multi-modal learning models for micro-video venue categorization, i.e., multi-task multimodal consistent learning model TRUMANN, tree-guided multimodal complementary learning strategy INTIMATE, and neural multimodal co-operative learning one NMCL. Specifically, the TRUMANN model is capable of learning a common feature space from multiple and heterogonous modalities and preserve the information of each modality via disagreement penalty. The INTIMATE co-regularizes the hierarchical smoothness and structure consistency within a unified model to learn the high-level sparse representations of micro-videos. Considering the timeliness and limited training samples, an online learning algorithm is developed to efficiently and incrementally strengthen the learning performance. And the NMCL model sheds light on characterizing and modeling the correlations between modalities, especially the consistent and complementary relations. In this model, we introduced a novel relation-aware attention mechanism to split the consistent information from the complementary one. Following that, we integrated the consistent information to learn an enhanced consistent vector and supplemented the complementary information to enrich this enhanced vector. And the experimental results on publicly accessible dataset have well validated the promising efficiency and effectiveness of these three models.

CHAPTER 5

Multimodal Transfer Learning in Micro-Video Analysis

5.1 BACKGROUND

In Chapter 4, we introduce a series of multimodal cooperative learning methods toward micro-video understanding, especially on the task of estimating their venue categories. It utilizes the consensus information on visual, acoustic, and textual modalities of micro-videos to recognize the venue information. This multi-modal method, however, overlooks that the description of each modality may differ dramatically. According to the experiments [191], the acoustic modality has the weakest capability of indicating venue information. It will cause the substantial cask effect, which is crucial to venue estimation and further content understanding.

To gain deep insights, we performed a user study to verify the acoustic influence on estimating venue category. Given 100 micro-videos randomly selected from Vine, 5 volunteers were blinded to the visual and textual modalities, tried to infer the venue category of each micro-video based on its acoustic modality, and subsequently rated the level of acoustic importance with one score from 1–5. As shown in Figure 5.1a, we have several observations: (1) the acoustic modality in 59% of micro-videos can benefit, more or less, the venue category estimation. It reveals the potential impact of the acoustic concepts. For example, recognizing *bird chirps* or *crowds cheering* from the acoustic modality is helpful for the estimation of *a park* or *concert*. However, (2) the acoustic modality in 84% of micro-videos is insufficiently descriptive (below 4) to support the venue category estimation, pertaining to their noise and low-quality. This study shows that detecting acoustic concepts is useful yet needs further enhancement.

5.2 RESEARCH PROBLEMS

Leveraging external rich sound knowledge to compensate the internal acoustic modality is an intuitive thought. It is, however, non-trivial owing to the following challenges: (1) as micro-videos often record events, it is desired to detect high-level acoustic concepts, which are more discriminative for event description [141]. We thus have to learn the conceptual representation of micro-videos; (2) micro-videos are about users' daily activities, which restricts us to harness the external real-life sound clips. However, to the best of our knowledge, there is no such sound data collection available. Moreover, (3) the external sounds are mono-modal data, whereas the micro-videos are unifying of textual, visual, and acoustic modalities. The micro-videos and external

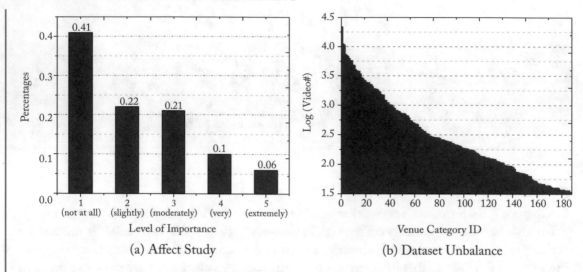

(a) Affect Study

(b) Dataset Unbalance

Figure 5.1: (a) Affect study regarding acoustic modality of micro-videos and (b) micro-video unbalance distribution over the dataset [191].

sounds are hence in two heterogeneous domains. How to integrate them within a unified model is largely untapped. Furthermore, according to our statistics over the benchmark dataset in [191], we observe a very severe problem of unbalanced training data, as shown in Figure 5.1b. This phenomenon tends to guide a bias *winner-takes-all* model, i.e., unseen samples are probably classified into the venue category with the most training micro-videos.

5.3   RELATED WORK

Our work is closely related to multimedia location estimation, dictionary learning, and acoustic concept detection. We have introduced multimedia location estimation and dictionary learning in Sections 4.3.1 and 4.3.3, respectively. We hence only detail the acoustic concept detection in this part.

Acoustic concept detection on the user-generated videos is a relatively new field in multimedia community [135], composed of the data-driven [17, 20] and task-driven [2, 128] approaches from the perspective of modeling acoustic concepts. The main motivation of acoustic concept detection is that audio analysis provides a complementary information to detect the specific events that are hardly identified with visual cues. Recent studies [167] have shown that detecting sound events to bridge the gap between the low-level features and the high-level semantics outperforms the pure feature-based approaches. Different from acoustic concept detection, we target at constructing a knowledge base of acoustic concepts and leveraging such base to strengthen the representation learning of micro-videos.

5.4   EXTERNAL SOUND DATASET

Micro-videos in Dataset II were collected from Vine and exclusively distributed in 442 venue categories. We filtered out those categories with less than 50 micro-videos to avoid the imbalanced classes. We ultimately left 270,145 micro-videos over 188 venue categories. Each micro-video is described by a rich set of features, namely, 4,096-d CNN visual features by AlexNet [82], 200-d Stacked Denosing Auto-encoder (SDA) acoustic features, and 100-d paragraph-to-vector textual features. Noticeably, in our selected dataset, 169 and 24,707 micro-videos do not have acoustic and textual modalities, respectively. We inferred their missing data via matrix factorization, which have been proven to be effective in the multi-modal data completion task [149].

As analyzed before, the acoustic modality is the least descriptive one and we expect to borrow the external sounds to enhance its discrimination. The scope of external sound dataset has direct effect on the performance of representation learning over micro-videos. Therefore, external sound construction is of importance. Indeed, there are several prior efforts on the sound clip collection. For example, Mesaros et al. [114] manually collected audio recordings from 10 acoustic environments and recognized them into 60 event-oriented concepts; Pancoast et al. [128] established 20 acoustic concepts relying on a small subset of TRECVID 2011; Burger et al. [12] extracted 42 concepts to describe distinct noise units from the soundtracks of 400 videos. We noticed that the existing external sound bases are either too small to cover the common acoustic concepts, or acquired from a narrow range of event-oriented videos. They are thus not feasible to our task.

To address this problem, we chose to collect sound clips from Freesound.[1] Freesound is a collaborative repository of Creative Commons licensed audio samples with more than 230,000 sounds and 4 million registered users as of February 2015. Short audio clips are uploaded to the website by its users, and cover a wide range of real-life subjects, like *applause* and *breathing*. Audio content in the repository can be tagged with acoustic concepts and browsed by standard text-based search. We first went through a rich set of micro-videos and manually defined 131 acoustic concepts, including the 60 acoustic concepts from the real-life recordings in [114]. Our pre-defined acoustic concepts are diverse and were treated as the initial seeds. We then fed these concepts into Freesound as queries to search the relevant sound clips. In this way, we gathered 16,363 clips. Each clip was manually labeled with several tags (i.e., acoustic concepts) by their owners and we in total obtained 146,580 acoustic concepts. To select the commonly heard acoustic concepts, we filtered out those concepts with less than 50 sound clips. Meanwhile, we adopted WordNet [78] to merge the acoustic concepts with similar semantic meanings, such as *kids* and *child*. Thereafter, we were left a set of 465 distinct acoustic concepts. Following that, we again fed each acoustic concept into Freesound as a query to acquire its sound clips with a number limit of 500. As a result, we gathered 45,948 sound clips. To ensure the quality of the sound data, we retrained acoustic concepts with at least 100 sound clips. We ultimately have 313 acoustic concepts and 43,868 sound clips. The statistics of the acoustic dataset are summarized in

---

[1]https://freesound.org/

Table 5.1. Some acoustic concept examples and their average sound durations are demonstrated in Figure 5.2. We can see that the external sound clips are very short. Similar to the micro-videos, the collected sound clips are also short and can be characterized by high-level concepts. Regarding each audio clip, we explored and extracted the same SDA acoustic features with those of the acoustic modality in micro-videos. We will clarify why we extracted this type of features in the experiments.

Table 5.1: Statistics of our collected external sound data

Concepts	Total Sound Clip	Sound Clips Per Concept	Average Duration	Average Concepts Per Sound Clip
313	43,868	140.15	14.99 s	2.99 (after data laundry)

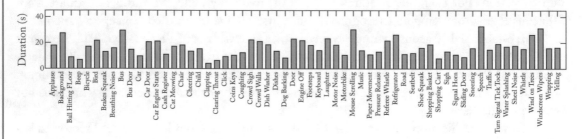

Figure 5.2: Exemplar demonstration of some acoustic concepts and their average sound clip durations.

5.5   DEEP MULTI-MODAL TRANSFER LEARNING

We formally introduce the problem definition. Suppose there are N micro-videos $\mathcal{X} = \{x_i\}_{i=1}^{N}$. For each micro-video $x \in \mathcal{X}$, we pre-segment it into three modalities $x = \{\mathbf{x}^v, \mathbf{x}^a, \mathbf{x}^t\}$, where-into the superscripts v, a, and t, respectively, represents the visual, acoustic, and textual one. To make more clear presentation, we denote $m \in \mathcal{M} = \{v, a, t\}$ as a modality indicator, and $\mathbf{x}^m \in \mathbb{R}^{D_m}$ as the D_m-dimensional feature vector over the m-th modality. And we associate $\mathbf{x}$ with one of the K pre-defined venue categories, namely an one-hot label vector $\mathbf{y}$. Our research objective is to generalize a venue estimation model over the training set to the new coming micro-videos.

To address these challenges, we first heuristically construct a set of 313 acoustic concepts covering most of the frequent real-life sounds, and collect 43,868 sound clips from Freesound. Consequently, we present a Deep trAnsfeR modEl, DARE for short, to effectively estimate the venue category. It jointly harnesses external sounds to strengthen the conceptual representation learning and regularizes the similarity preservation to alleviate the problem of unbalanced training samples. The scheme of our proposed DARE approach is illustrated in Figure 5.3. To be more specific, we first segment each micro-video into visual, textual, and acoustic modalities. We then project the low-level representations of each modality to conceptual ones by distinct mapping functions. To transfer the external sound knowledge to strengthen the acoustic modality, we force them to share the same low-level feature space and mapping function. Following that, we propose a deep multi-modal fusion method, which utilizes complementary information from each modality, encodes the category structure information by similarity preservation, and uncover the nonlinear correlations between concepts. We ultimately feed the fused representations into a prediction function to estimate the venue categories.

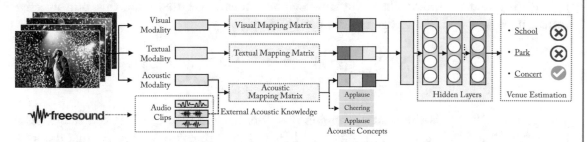

Figure 5.3: Schematic illustration of our proposed deep transfer model. It transfers knowledge from external sound clips to strengthen the description of the internal acoustic modality in micro-videos. Meanwhile, it conducts a deep multi-modal fusion toward venue category estimation.

5.5.1   SOUND KNOWLEDGE TRANSFER

In order to leverage the external sound knowledge to enhance the acoustic modality in micro-videos, we have two assumptions: (1) concept-level representations are more discriminative to characterize each modality in micro-videos and the external sounds; and (2) the natural correlation between the acoustic modality in micro-videos and the real-life sounds motivate us to assume that they share the same acoustic concept space.

As to the concept-level representation, one intuitive thought is multi-modal dictionary learning, whereby the atoms in the dictionaries are treated as concepts. We, however, argue that the implicit assumption of multi-modal dictionary learning does not always hold in some real-world scenarios: the dictionaries of distinct modalities share the same concept space. Considering the micro-video analysis as an example, the acoustic modality may contain the concept of *chirp of birds* that is hardly expressed by the visual modality. In the textual one, it may signal some atoms related to *sense of smell*, which also impossibly appear in the visual modality. Therefore, it is not necessary to enforce the dictionaries of different modalities to contain the same set of concepts. To avoid such a problem, we propose to learn a separate mapping function for each modality that is able to project the low-level features to concept-level representations. Analogous to the dictionaries in dictionary learning paradigms, the mapping functions are the concept-feature distributions.

Let $\widetilde{\mathcal{X}}^a = \{\widetilde{\mathbf{x}}_i^a\}_{i=1}^{N'}$ be the dataset of external sounds. These sounds share the same low-level feature space with the acoustic modality in micro-videos (i.e., $\widetilde{\mathbf{x}}^a \in \mathbb{R}^{D_a}$). For each sound clip $\widetilde{\mathbf{x}}^a$, we denote its corresponding concept-wise representation as $\widetilde{\mathbf{a}}^a \in \mathbb{R}^{K'}$ over K' acoustic concepts, whereby K' equals to the number of acoustic concepts in this work, i.e., 313. It is worth noting that $\widetilde{\mathbf{a}}^a$ is observable, since we know the associated tags (acoustic concepts of each collected sound clip). During learning, we aim to use the concept space of the external real-life sounds to represent the acoustic modality in the given micro-video. This is accomplished by ensuring that $\mathbf{x}^a$ and $\widetilde{\mathbf{x}}^a$ share the same mapping function. Based upon this, our objective function $\mathcal{J}_1$ of sound knowledge transfer can be stated as

$$\mathcal{J}_1 = \frac{1}{N} \sum_{\mathbf{x} \in \mathcal{X}} \sum_{m \in \mathcal{M}} \|\mathbf{D}^m \mathbf{x}^m - \mathbf{a}^m\|^2 + \frac{1}{N'} \sum_{\mathbf{x} \in \widetilde{\mathcal{X}}} \|\mathbf{D}^a \widetilde{\mathbf{x}}^a - \widetilde{\mathbf{a}}^a\|^2, \tag{5.1}$$

where $\mathbf{D}^a \in \mathbb{R}^{D_a \times K'}$ is the shared mapping function, bridging the gap between the external sounds and the internal acoustic modality, whereinto its i-th column $\mathbf{d}_i^a$ represents the low-level feature for the i-th concept, such as *footsteps* or *clearing throat*; and $\mathbf{a}^a \in \mathbb{R}^{K'}$ is the desired concept-level representation of $\mathbf{x}$ over the K' acoustic concepts; and $\mathbf{D}^v$ and $\mathbf{a}^v$ ($\mathbf{D}^t$ and $\mathbf{a}^t$) are analogous to $\mathbf{D}^a$ and $\mathbf{a}^a$. Noticeably, $\mathbf{D}^v$ is an identity matrix, slightly different from other two mapping functions, since the visual features are sufficiently abstractive extracted by AlexNet.

5.5.2 MULTI-MODAL FUSION

As aforementioned, multi-modalities provide complementary cues. We thus argue that multi-modal fusion can provide comprehensive and informative description for micro-videos. In our case, we adopt early fusion strategy for simplicity. Formally, for each micro-video $\mathbf{x}$, we concatenate $\mathbf{a}^v$, $\mathbf{a}^a$, and $\mathbf{a}^t$ into one vector as

$$\mathbf{a} = [\mathbf{a}^v, \mathbf{a}^a, \mathbf{a}^t], \tag{5.2}$$

where $\mathbf{a} \in \mathbb{R}^{D_v + K' + D_t}$ is the desired multi-modal representation for $\mathbf{x}$, whereinto $\mathbf{a}^v$, $\mathbf{a}^a$, and $\mathbf{a}^t$, respectively, denote the concept-level representation over the visual, acoustic, and textual modalities.

To alleviate the problem of unbalanced training samples, we further regularize $\mathbf{a}_i$ for each micro-video $\mathbf{x}_i$ by similarity preservation. In particular, if two micro-videos are in the same venue category, they should have similar representations in the latent space. Otherwise, they have dissimilar ones. This suits well the paradigm of graph embedding [182], which injects the label information into the embeddings.

Formally, we denote $(\mathbf{x}_i, \mathbf{x}_j)$ as the pair of the i-th and j-th samples, and define a pairwise class indicator as

$$\gamma_{ij} = \begin{cases} +1, & \text{if } \mathbf{x}_i \text{ and } \mathbf{x}_j \text{ are with the same label;} \\ -1, & \text{otherwise.} \end{cases} \tag{5.3}$$

To encode the similarity preservation, we minimize the cross entropy loss of classifying all the pairs into a label γ,

$$\sum_{i,j=1}^{N} -\mathbb{I}\left(\gamma_{ij} = 1\right) \log \sigma \left(\mathbf{a}_i^\top \mathbf{a}_j\right) - \mathbb{I}\left(\gamma_{ij} = -1\right) \log \sigma \left(-\mathbf{a}_i^\top \mathbf{a}_j\right), \tag{5.4}$$

where $\mathbb{I}(\cdot)$ is a binary indicator function that outputs 1 when the argument is true, otherwise 0; and $\sigma(\cdot)$ is the sigmoid function. We can equivalently rewrite the above equation as

$$\mathcal{J}_2 = -\sum_{i=1}^{N} \sum_{j=1}^{N} \log \sigma \left(\gamma_{ij} \mathbf{a}_i^\top \mathbf{a}_j\right). \tag{5.5}$$

It is very time-consuming to directly optimize Eq. (5.5) due to the huge amount of the instance pairs, i.e., $O(N^2)$ w.r.t N samples.

To reduce the computing load, we turn to negative sampling [115]. In particular, for a given micro-video sample $\mathbf{x}$, we respectively sampled S positive and S negative micro-videos from $\mathbf{x}$'s own category and its non-categories following a distribution $(\mathbf{x}_i, \mathbf{x}_j, \gamma_{ij})$. Formally, we uniformly sample the first instance $\mathbf{x}_i$. We then sample the next instance $\mathbf{x}_j$ with a probability s_{ij} that represents the geometric closeness between $\mathbf{x}_i$ and $\mathbf{x}_j$. We calculate s_{ij} with the radial basis function kernel as

$$s_{ij} = \frac{1}{|\mathcal{M}|} \sum_{m \in \mathcal{M}} \exp \left(-\frac{\left\| \mathbf{x}_i^m - \mathbf{x}_j^m \right\|^2}{\delta_m^2} \right), \tag{5.6}$$

where δ_m^2 is a radius parameter that is set as the median of the Euclidean distances of all samples on the modality m.

5.5.3 DEEP NETWORK FOR VENUE ESTIMATION

After obtaining the multi-modal representations, we add a stack of fully connected layers, which enables us to capture the nonlinear and complex interactions between the visual, acoustic, and textual concepts. More formally, we define these fully connected layers as

$$
\begin{cases}
\mathbf{e}_1 = \sigma_1 \left(\mathbf{W}_1 \mathbf{a} + \mathbf{b}_1 \right), \\
\mathbf{e}_2 = \sigma_2 \left(\mathbf{W}_2 \mathbf{e}_1 + \mathbf{b}_2 \right), \\
\cdots\cdots \\
\mathbf{e}_L = \sigma_L \left(\mathbf{W}_L \mathbf{e}_{L-1} + \mathbf{b}_L \right),
\end{cases}
\tag{5.7}
$$

where $\mathbf{W}_l$, $\mathbf{b}_l$, σ_l, and $\mathbf{e}_l$ denote the weight matrix, bias vector, activation function, and output vector in the l-th hidden layers, respectively. As for activation function in each hidden layer, we choose Rectifier (ReLU) to learn higher-order concept interactions in a nonlinear way. Regarding the size of hidden layers, common solutions are following the tower, constant, and diamond patterns.

The output of the penultimate hidden layer is flattened to a dense vector $\mathbf{e}_L$, which is passed to a fully connected softmax layer. It computes the probability distributions over the venue category labels as

$$
p \left(\widehat{y}_k | \mathbf{e}_L \right) = \frac{\exp \left(\mathbf{e}_L^\top \mathbf{w}_k \right)}{\sum_{k'=1}^{K} \exp \left(\mathbf{e}_L^\top \mathbf{w}_{k'} \right)},
\tag{5.8}
$$

where $\mathbf{w}_k$ is a weight vector of the k-th venue category and $\mathbf{e}_L$ can be viewed as the final abstract representation of the input $\mathbf{x}$. Thereafter, we obtain the probabilistic label vector $\widehat{\mathbf{y}} = [\widehat{y}_1, \ldots, \widehat{y}_K]$ over the K venue categories.

Thereafter, we adopt the regression-based function to minimize the loss between the estimated label vector and its target values, as

$$
\mathcal{J}_3 = \frac{1}{2} \sum_{\mathbf{x} \in \mathcal{X}} \| \mathbf{y} - \widehat{\mathbf{y}} \|^2,
\tag{5.9}
$$

where an ideal model should predict the venue category correctly for each micro-video.

We ultimately obtain our objective function of the proposed deep transfer model by jointly regularizing the sound knowledge transfer, multi-modal fusion, and DNN for venue estimation as

$$
\mathcal{J} = \mathcal{J}_1 + \mathcal{J}_2 + \mathcal{J}_3.
\tag{5.10}
$$

5.5.4   TRAINING

We adopted the stochastic gradient descent to train our model in a mini-batch mode and updated the corresponding model parameters using back propagation. In particular, we first sampled a batch of instances and took a gradient step to optimize the loss function of external sound transfer. We then sampled a batch of $(\mathbf{x}_i, \mathbf{x}_j, \gamma_{ij})$ and took another gradient step to optimize the loss of multi-modal embedding learning. Thereafter, we optimized the loss function of venue category estimation. To speed up the convergence rate of SGD, various modifications to the update rule have been explored, namely, momentum, adagrad, and adadelta.

While DNNs are powerful in representation learning, a deep architecture easily leads to the overfitting on the limited training data. To remedy the overfitting issue, we conducted dropout to improve the regularization of our deep model. The idea is to randomly drop part of neurons during training. As such, dropout acts as an approximate model averaging. In particular, we randomly dropped ρ of $\mathbf{a}$, whereinto ρ is the dropout ratio. Analogously, we also conducted dropout on each hidden layer.

5.6 EXPERIMENTS

To thoroughly justify the effectiveness of our proposed deep transfer model, we carried out extensive experiments over Dataset II to answer the following research questions.

- **RQ1:** Are the extracted 200-D SDA features discriminative to represent the external sounds?

- **RQ2:** Can our DARE approach outperform the state-of-the-art baselines for micro-video categorization?

- **RQ3:** Is the external sound knowledge helpful for boosting the categorization accuracy and does the external data size affect the final results?

- **RQ4:** Does the proposed DARE model converge and does different parameter settings affect the final results?

5.6.1 EXPERIMENTAL SETTINGS

We divided our dataset into three parts: 132,370 for training, 56,731 for validation, and 81,044 for testing. The training set was used to adjust the parameters, while the validation one was used to verify that any performance increase over the training dataset actually yields an accuracy increase over a dataset that has not been shown to the model before. The testing set was used only for testing the final solution to confirm the actual predictive power of our model with optimal parameters.

Baselines
We chose the following methods as our baselines.

- **Default:** For any given micro-video, we dropped it into the category with maximum micro-videos by default.

- **D³L:** Data-driven dictionary learning is a classic mono-modal unsupervised dictionary learning framework utilizing elastic-net proposed by Culotta et al. [35]. A late fusion by the softmax model over the learned sparse representations is incorporated.

- **MDL:** This baseline is the traditional unsupervised multi-modal dictionary learning [117]. It is also followed by a late fusion with softmax.

- **MTDL:** This tulti-modal task-driven dictionary learning approach [6] learned the discriminative multi-modal dictionaries simultaneously with the corresponding venue category classifiers.

- **TRUMANN:** This is a tree-guided multi-task multi-modal learning method, which considers the hierarchical relatedness among the venue categories.

- **AlexNet:** In addition to the aforementioned shallow learning methods, we added four deep models into our baseline pool, i.e., the AlexNet model with zero, one, two, and three hidden layers, whereby their inputs are the original feature concatenation of three modalities and they predict the final results with a softmax function.

Indeed, our model is also related to transfer learning methods. However, existing transfer models [43, 105, 185] are not suitable to our task, since they work by leveraging one source domain to support one target domain. Yet, our task has one source domain (external sounds) and three target domains (three modalities). Therefore, we did not compare our method with existing transfer learning methods.

Parameter Settings

We implemented our DARE model with the help of Tensorflow.[2] To be more specific, we randomly initialized the model parameters with a Gaussian distribution for all the deep models in this chapter, whereby we set the mean and standard derivation as 0 and 1, respectively. The mini-batch size and learning rate for all models was searched in [256, 512, 1,024] and [0.0001, 0.0005, 0.001, 0.005, 0.1], respectively. We selected Adagrad as the optimizer. Moreover, we selected the constant structure of hidden layers, empirically set the size of each hidden layer as 1,024 and the activation function as ReLU. For our DARE, we set the embedding sizes of visual, acoustic, and textual mapping matrices as 4,096, 313, and 200, respectively, which can be treated as the extra hidden layer for each modality. Without special mention, we employed one hidden layer and one prediction layer for all the deep methods. We randomly generated five different initializations and fed them into our DARE. For other competitors, the initialization procedure is analogous to ensure the fair comparison. We reported the average testing results over five round results and performed paired t-test between our model and each of baselines over five-round results.

---

[2]https://www.tensorflow.org

5.6.2   ACOUSTIC REPRESENTATION (RQ1)

To represent each external sound clip, we first extracted two kinds of commonly used features, i.e., 513-d spectrum and 39-d MFCCs, with a 46-ms window size and 50% overlap via librosa.[3] We then employed the mean- and max-pooling strategy to represent each clip. Besides, we also adopted theano[4] to learn a 200-d SDA feature vector of each clip, whose input is the concatenated feature vector of 513-d spectrum (mean), 513-d spectrum (max), 39-d MFCCs (mean), and 39-d MFCCs (max).

In order to justify the representativeness of the extracted features on the external sounds, we, respectively, fed the features into a softmax model to learn a sound clip classifier. In particular, we treated each acoustic concept as a label. We performed a 10-fold cross-validation. The results are summarized in Table 5.2. We can see that the SDA features are the most discriminant one. We conducted significance test between SDA and each of the others regarding macro-F1 based on the 10-round results. All the p-values are greatly smaller than 0.05, which indicates that SDA is statistically significant better. That is why we used the SDA feature in the hereafter experiments.

Table 5.2: Discrimination comparison among different acoustic features on Dataset II (p-value*: p-value over macro-F1)

Feature Sets	Macro-$F1$	Micro-$F1$	P-value*
Spectrum (mean)	6.23 ± 0.30%	8.87 ± 0.17%	7.4e-6
Spectrum (max)	5.88 ± 0.49%	8.25 ± 0.53%	4.8e-6
MFCC (mean)	4.92 ± 0.63%	11.36 ± 0.96%	3.6e-4
MFCC (max)	9.21 ± 0.46%	15.72 ± 0.77%	4.2e-3
SDA	**12.74 ± 0.62%**	**17.09 ± 0.69%**	–

5.6.3   PERFORMANCE COMPARISON (RQ2)

We summarized the performance comparison among all the methods in Table 5.3. We have the following observations.

- As expected, default achieves the worst performance, especially w.r.t. macro-F1.

- In terms of micro-F1, performance of three dictionary learning baselines is comparative; whereas D^3L achieves the worst macro-F1. This may be due to that D^3L does not differ the three modalities.

---

[3]http://librosa.github.io/librosa
[4]http://deeplearning.net/software/theano

Table 5.3: Performance comparison between our model and the baselines on Dataset II (p-value1* and p-value2* are, respectively, p-value over micro-F1 and macro-F1)

Feature Sets	Micro-F1	Macro-F1	P-value1*	P-value2*
Default	11.40%	0.53%	$1.93e\text{-}9$	$1.41e\text{-}8$
MDL	20.46 ± 0.49%	7.06 ± 0.27%	$3.39e\text{-}8$	$2.01e\text{-}7$
D^3L	19.03 ± 0.29%	3.87 ± 0.24%	$1.29e\text{-}8$	$2.29e\text{-}8$
MTDL	20.67 ± 0.29%	6.16 ± 0.24%	$4.29e\text{-}8$	$1.94e\text{-}8$
$AlexNet_0$	25.95 ± 0.08%	6.04 ± 0.07%	$9.81e\text{-}7$	$1.36e\text{-}8$
$AlexNet_1$	28.95 ± 0.17%	9.45 ± 0.13%	$2.15e\text{-}5$	$1.38e\text{-}7$
$AlexNet_2$	29.04 ± 0.17%	10.86 ± 0.18%	$4.02e\text{-}5$	$1.24e\text{-}6$
$AlexNet_3$	28.55 ± 0.49%	10.65 ± 0.34%	$1.91e\text{-}4$	$4.87e\text{-}6$
TRUMANN	25.27 ± 0.17%	5.21 ± 0.29%	$2.46e\text{-}7$	$9.23e\text{-}8$
Our DARE	**31.21 ± 0.22%**	**16.66 ± 0.30%**	–	–

- The TRUMANN model is better than dictionary learning methods, since it considers the hierarchical structure of venue categories.

- AlexNet with at least one hidden layer remarkably outperforms $AlexNet_0$ and dictionary learning ones across metrics. This demonstrates the advantage of deep models.

- Among the AlexNet series, it is not the deeper the better. This is caused by the intrinsic limitation of AlexNet. (We will detail it in RQ4.)

- Without a doubt, our proposed model achieves the best regardless of the metrics. This justifies the effectiveness of our model. From the perspective of macro-F1, our model makes noteworthy progress. This further shows the rationality of similarity preservation by encoding the structural category information. In addition, we also conducted pair-wise significant test between our model and each baseline. All the p-values are greatly smaller than 0.05, which indicates the performance improvement is statistically significant.

5.6.4   EXTERNAL KNOWLEDGE EFFECT (RQ3)

We carried out experiments to study the effect of external sound knowledge on our model. In particular, we varied the number of external acoustic concepts from 0 to 313. Figures 5.4a and 5.4b illustrate the performance of our model according to the external data size w.r.t macro-F1 and micro-F1, respectively. It is clear that these two curves go up very fast. Such phenomenons tell us that transferring external sound knowledge is useful to boost the categorization accuracy. Also, it signals that the more external sounds are involved, the better performance we

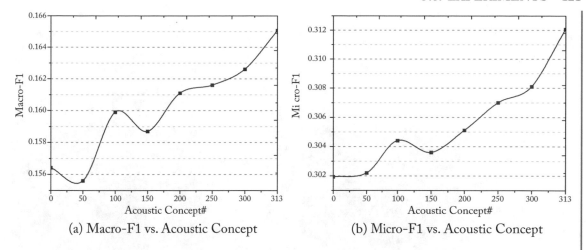

(a) Macro-F1 vs. Acoustic Concept   (b) Micro-F1 vs. Acoustic Concept

Figure 5.4: Performance of our model w.r.t. the number of external acoustic concepts.

will achieve, since it is able to cover a much wider range of acoustic concepts appeared in micro-videos.

5.6.5   VISUALIZATION

We conducted experiments to shed some light on the correlation between venue categories and acoustic concepts. In particular, we calculated the correlations between acoustic concepts and venue categories via producing inner products on the conceptual distributions and venue label vectors of samples. To save the space, we visualized part of correlation matrix via a heat map, where lighter color indicates weak correlation and vice versa, as shown in Figure 5.5. We can see that almost each selected venue category is tightly related to several acoustic concepts. Moreover, different venues emphasize a variety of acoustic concepts. For example, the micro-videos with venue of *Italian Restaurant* and *College (University)* have significant correlations with the onomatopoeia concepts, such as *rattle, jingle*, and *rumble*; meanwhile, several motion concepts, such as *scream, running*, and *clapping* provide clear cues to infer the venue information of *Housing Development, Gym*, and *Playground*, respectively. These observations agree with our daily experiences and further demonstrate the potential influence of acoustic information on the task of venue category estimation.

5.6.6   STUDY OF DARE MODEL (RQ4)

We wonder whether our model converges and how fast it is. To answer this question, we plot the training loss, macro-F1, and micro-F1 with respect to the number of iterations in Figures 5.6a, 5.6b, and 5.6c, respectively. From these three sub-figures, it can be seen that the training loss of our proposed DARE model decreases quickly within the first 10 iterations, and accordingly

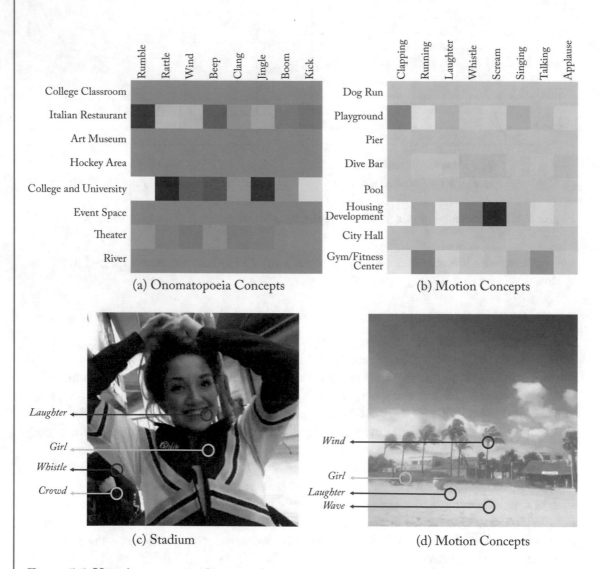

Figure 5.5: Visualization regarding correlations between venue category and two types of acoustic concepts.

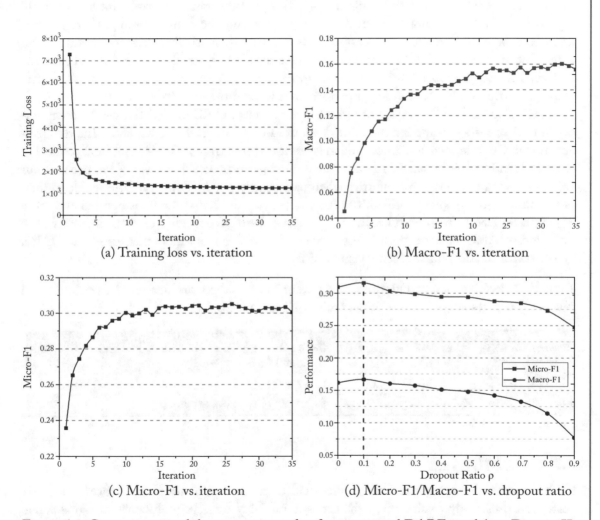

Figure 5.6: Convergence and dropout ratio study of our proposed DARE model on Dataset II.

the performance is also boosted very fast. This demonstrates the rationality of a learning model. In addition, the loss and performance tend to be stable at around 30 iterations. This signals the convergence property of our model and also indicates its efficiency.

The key idea of dropout technique is to randomly drop units (along with their connections) from the neural network during training. This prevents units from co-adapting too much. Figure 5.6d displays the macro-F1 and micro-F1 by varying the dropout ratio ρ. From this figure, it can be seen that the two measurements consistently reach their best value when using a dropout ratio of 0.1. After 0.1, the performance decreases gradually as the dropout ratio increases. This may be caused by insufficient information. Also, we can see that our model suffers from overfitting with relatively lower performance when dropout ratio is set as 0.

We also studied the impact of hidden layers on our DARE model. To save the computational tuning costs, we applied the same dropout ratio 0.1 for each hidden layer. The results of our model with one, two, and three hidden layers are summarized in Table 5.4. Usually, stacking more hidden layers is beneficial to boost the desired performance. However, we notice that our model achieves the best across metrics when having only one hidden layer. This is due to that, as the authors of AlexNet clarified, the current 7-layer AlexNet structure is optimal and more layers would lead to worse results. In our work, the abstractive features of visual modality were extracted by AlexNet with seven layers. Therefore, stacking more hidden layers in our DARE model seems to add more hidden layers to AlexNet.

Table 5.4: Performance of DARE with different hidden layers on Dataset II (p-value1* and p-value2* are, respectively, p-value over micro-F1 and macro-F1)

Hidden Layers	Micro-F1	Macro-F1	P-value1*	P-value2*
[1024]	31.21 ± 0.22%	16.66 ± 0.30%	–	–
[1024, 1024]	30.67 ± 0.06%	15.57 ± 0.03%	$1.32e\text{-}2$	$3.50e\text{-}3$
[1024, 1024, 1024]	29.43 ± 0.02%	13.37 ± 0.04%	$1.17e\text{-}4$	$1.57e\text{-}6$

5.7 SUMMARY

In this chapter, we study the task of micro-video category estimation. In particular, we first perform a user study to show that the acoustic modality conveys useful cues to signal venue information, yet it is of low-quality. We then point out that the training sample distribution over venue categories are extremely unbalanced. To address these problems, we present a deep transfer model, which is able to transfer external sound knowledge to strengthen the low-quality acoustic modality in micro-videos, and also alleviate the problem of unbalanced training samples via encoding the category structure information. To justify our model, we constructed the external sound sets with diverse acoustic concepts, and released it to facilitate other researchers. Experimental results on a public benchmark micro-video dataset well validate our model.

CHAPTER 6

Multimodal Sequential Learning for Micro-Video Recommendation

6.1 BACKGROUND

As the micro-videos surge, it becomes increasingly difficult and expensive for users to locate their desired micro-videos from the vast candidates. In light of this, it is crucial to build a personalized recommendation system to intelligently route micro-videos to the target users.

6.2 RESEARCH PROBLEMS

Building a personalized recommendation system for micro-video services is non-trivial, due to the following reasons. (1) **Diverse and dynamic interest**. On the one hand, users' interest evolves over time, and it is hence a sequential expression. For example, as shown in Figure 6.1, a user likes cooking videos at time t_1, but may prefer dance videos at t_2. On the other hand, users' interest is diverse, namely a user may be fond of multiple topics at the same time. In a sense, personalized recommendation requires to simultaneously model users' dynamic and diverse interest information. (2) **Multi-level interest**. Users may have different interaction types on micro-videos, including "click," "like," and "follow," which signal different degrees of interest. For example, "click" means the user is attracted to the micro-video, "like" is one much enjoys and appreciates the micro-video, and "follow" refers to the user likes the micro-video very much and wishes to see it again in future. Heretofore, how to integrate the various degrees of interest into personalized recommendation is largely untapped. (3) **True negative samples**. As we know, prior methods commonly assume that nonpositive items are negative samples, which is hardly reliable to infer which item a user did not like. Different from these models, we are able to obtain true negative samples, i.e., micro-videos that users preview the thumbnails yet no "click" occurs. Therefore, how to utilize these true negative samples to explicitly model users' uninterested information becomes a crucial problem.

For the past few years, several studies have been conducted on the personalized recommendation, such as collaborative filtering based models [7, 25, 69], content-based systems [34, 113, 129, 206, 209], and hybrid methods [48, 201]. Although these methods produce promising performance on recommendation, most of them suppose users' interest as static. In-

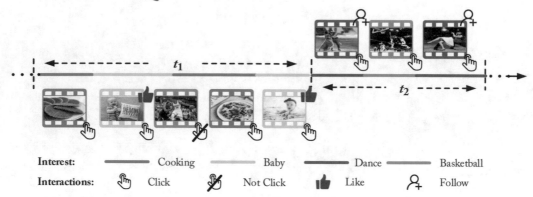

Figure 6.1: Illustration of a user's historical interactions with micro-videos, which reflects the user's diverse, dynamic, and multi-level interest.

spired by this, some researches consider users' interest as dynamic when designing recommendation systems and have achieved better performance [28, 52, 132, 152, 157, 203]. They, however, overlook the diverse and multi-level characteristics of users' interest. Moreover, all the aforementioned methods commonly assume that items not been clicked by users are negative and utilize them [60, 127, 137] or sample part of them as negative samples [61, 66] to represent users' uninterested items. However, these presumed negative samples may be not truly negative, and they hence may confuse the recommendation system. As we can see, the existing studies neither consider the diverse and multi-level interest nor exploit users' true uninterested items to model the recommendation system. Therefore, they cannot be directly applied to the micro-video recommendation.

6.3   RELATED WORK

Recommender systems are vital in video communities, such as Youtube, Vimeo,[1] and Veoh.[2] The exiting methods can be roughly categorized as collaborative filtering-based methods [7, 25, 69], content-based methods [34, 113, 129, 206, 209], and hybrid methods [48, 201]. In terms of collaborative filtering, Baluja et al. [7] utilized the random walk through a co-view graph to recommend YouTube videos. Chen et al. [25] integrated an attention mechanism into collaborative filtering with implicit feedback and evaluated its effectiveness in multimedia recommendation. However, collaborative filtering based methods cannot well solve the cold start problem. By contrast, the content-based methods recommend videos by calculating the similarity between new videos and users' historical accessed videos. For example, Mei et al. [113] proposed a contextual recommendation system based on multimodal fusion and relevance feedback. With respect

[1]https://vimeo.com/
[2]https://www.veoh.com/

to the hybrid models, they aim to combine the above two methods within a unified framework. For example, the recommendation model presented in [201] generates multiple ranking lists via exploring different information sources in a multi-task framework. Since the underlying assumption of the traditional video recommendation models is that users' interest is static, therefore they cannot be applied to extract users' dynamic interest.

Recently, many models have been proposed to characterize users' dynamic preferences. These methods are in three variants: CNN-based methods [152, 157], recurrent neural network (RNN) based methods [52, 132], and self-attention based methods [28, 203]. As a typical example in the first category, Tuan et al. [157] utilized 3-D CNNs to combine session clicks and content features to generate recommendations. As for RNN based methods, Quadrana et al. [132] proposed the RNN based approach for session-based recommendation, which relays and evolves latent hidden states of the RNNs across user sessions. In [52], the authors proposed a dynamic RNN to model users' dynamic interest for the personalized video recommendation. Due to the high time consumption and long sequence restriction, the self-attention mechanism has been applied to recommender systems and gained impressive performance. For example, Zhou et al. [203] proposed an attention-based user behavior model by considering heterogeneous user behaviors in e-commerce. Although the aforementioned methods have considered users' dynamic interest and been successfully applied to video communities, they are inadequate to handle micro-video communities due to their different characteristics. In particular, micro-video communities continuously route micro-videos to users and users click their interested ones by previewing the thumbnails, whereas traditional video communities are apt to display users' interested videos via their query information. In addition, users' interest information in micro-video communities has a multi-level structure.

6.4   MULTIMODAL SEQUENTIAL LEARNING

To address the aforementioned problems, in this chapter, we develop an end-to-end temporAL graPh-guIded recommeNdation systEm, dubbed ALPINE, to route micro-videos. The scheme of our proposed approach is illustrated in Figure 6.2. Specifically, to model users' diverse and dynamic interest, we encode users' click history information into a graph where the node refers to micro-videos in the click history and the edge between two nodes stands for the temporal relationship. Based upon this graph, we design a novel long short-term memory (LSTM) network to learn users' interest representation. Afterward, we estimate the click probability via calculating the similarity between the users' interest representation and the embedding of the given micro-video. Considering that users' interest is multi-level, we introduce a user matrix to enhance the user interest modeling by incorporating their "like" and "follow" information. And at this step, we also get a click probability with respect to users' more precise interest information. Analogously, since we know the sequence of users' disliked micro-videos, another temporal graph-based LSTM is built to characterize users' uninterested information, and the other click probability can be estimated based on true negative samples. We can thus obtain a click prob-

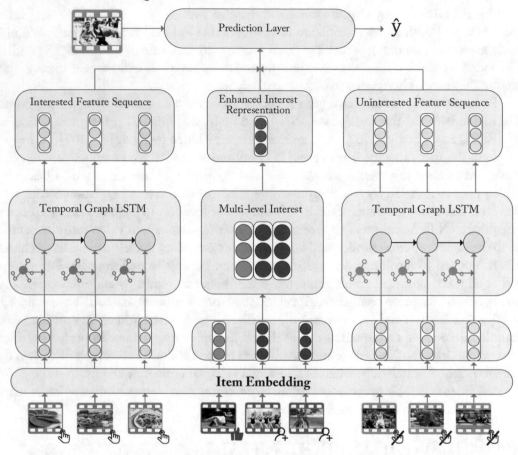

Figure 6.2: Illustration of our proposed ALPINE model.

ability regarding users' uninterested information. Finally, the weighted sum of the above three probability scores is set as our final prediction result.

Let v and u denote a micro-video and a user, respectively. We present the user's historical information as a sequence of micro-videos $\mathcal{U} = \{(u, v_j^t)\}_{t=1}^m$, where $j \in \{c, n, l, f\}$, respectively, represents user's "click," "not click," "like," and "follow" behaviors, and m is the length of the sequence. As the user's interest is multi-level, its sequential behaviors can be segmented into four sub-sequences, namely "click" sequence $\mathcal{U}_c = \{(u, v_c^{t_c})\}_{t_c=1}^{m_c}$, "not click" sequence $\mathcal{U}_n = \{(u, v_n^{t_n},)\}_{t_n=1}^{m_n}$, "like" sequence $\mathcal{U}_l = \{(u, v_l^{t_l})\}_{t_l=1}^{m_l}$, and "follow" sequence $\mathcal{U}_f = \{(u, v_f^{t_f})\}_{t_f=1}^{m_f}$, where $m_c + m_n + m_l + m_f = m$. As such, the micro-video recommendation problem can be formally defined as:

Input: The user's multi-level behavior sequences $\mathcal{U}_c$, $\mathcal{U}_n$, $\mathcal{U}_l$, and $\mathcal{U}_f$, and the given micro-video v_{new}.

Output: A recommendation system predicting the click probability of the user u on the new micro-video v_{new}.

6.4.1 THE TEMPORAL GRAPH-BASED LSTM LAYER

To model users' dynamic interest from their historical "click" information $\mathcal{U}_c$, a direct way is to utilize the LSTM network to model the temporal sequence and obtain their interest representation. Formally, the above process can be formulated as

$$
\begin{cases}
\mathbf{i}_t = & \sigma \left(\mathbf{W}_{ix}\mathbf{x}_t + \mathbf{W}_{ih}\mathbf{h}_{t-1} + \mathbf{b}_i \right), \\
\mathbf{f}_t = & \sigma \left(\mathbf{W}_{fx}\mathbf{x}_t + \mathbf{W}_{fh}\mathbf{h}_{t-1} + \mathbf{b}_f \right), \\
\mathbf{o}_t = & \sigma \left(\mathbf{W}_{ox}\mathbf{x}_t + \mathbf{W}_{oh}\mathbf{h}_{t-1} + \mathbf{b}_o \right), \\
\mathbf{u}_t = & \tanh \left(\mathbf{W}_{ux}\mathbf{x}_t + \mathbf{W}_{uh}\mathbf{h}_{t-1} + \mathbf{b}_u \right), \\
\mathbf{c}_t = & \mathbf{i}_t \odot \mathbf{u}_t + \mathbf{f}_t \odot \mathbf{c}_{t-1}, \\
\mathbf{h}_t = & \mathbf{o}_t \odot \tanh \left(\mathbf{c}_t \right),
\end{cases}
\tag{6.1}
$$

where $\mathbf{x}_t$ is the micro-video embedding at the time step t; $\mathbf{i}_t$, $\mathbf{f}_t$, $\mathbf{o}_t$, $\mathbf{c}_t$, and $\mathbf{h}_t$, respectively, denote the input gate, forget gate, output gate, memory cell, and hidden state; σ denotes the logistic sigmoid function; and $\odot$ denotes element wise multiplication. Although the LSTM network is capable of memorizing information from sequence data, we argue that it is insufficient to capture user's diverse interest from the very long historical sequence. Particularly, if the user's historical sequence only relates to one topic, the LSTM network indeed can capture user's single interest. However, as discussed before, interest is diverse, as shown in Figure 6.1. Thereby, it may fail to memorize the user's diverse interest information from the very long sequence.

To tackle the aforementioned problem, we consider to enhance the memorization of user's diverse interest by integrating an interest graph into the LSTM network. The detail of our temporal graph-based LSTM layer is illustrated in Figure 6.3. In particular, given the user's click sequence $\mathcal{U}_c$, we build a temporal graph $\mathcal{G}_c =< v_c, e_c >$. We view micro-videos in $\mathcal{U}_c$ as nodes, and link two nodes according to the following two rules: (1) to model the user's dynamic interest, each micro-video $v_c^{t_c}$ should be connected with its preceding micro-video $v_c^{t_c-1}$, namely $< v_c^{t_c-1}, v_c^{t_c} >$; and (2) to memorize the diverse interest of the user, we force each micro-video to link with the preceding micro-videos which share the similar visual information. Given a micro-video $v_c^{t_c}$, we estimate its similarity with respect to its pre-context micro-videos and connect it with the most similar one,[3] namely $< v_c^{t_c^*}, v_c^{t_c} >$. In light of this, we construct a temporal interest graph, where each node represents one of the user's interested micro-videos, and each edge

---

[3]The number of similar micro-videos that should be linked is detailed in Section 4.6.

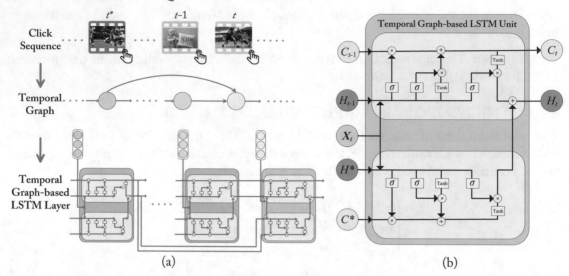

Figure 6.3: Structure of the temporal graph-based LSTM layer: (a) illustration of the temporal graph construction and (b) details of the temporal graph-based LSTM unit.

represents the relationship between the user's interested micro-videos. Moreover, for extracting the user's interested feature sequence, we design a novel graph-based LSTM network. Formally, we formulate this network as follows:

$$
\begin{cases}
\mathbf{i}_t = & \sigma\left(\mathbf{W}_{ix}\mathbf{x}_t + \mathbf{W}_{ih}\mathbf{h}_{t-1} + \mathbf{b}_i\right), \\
\mathbf{f}_t = & \sigma\left(\mathbf{W}_{fx}\mathbf{x}_t + \mathbf{W}_{fh}\mathbf{h}_{t-1} + \mathbf{b}_f\right), \\
\mathbf{o}_t = & \sigma\left(\mathbf{W}_{ox}\mathbf{x}_t + \mathbf{W}_{oh}\mathbf{h}_{t-1} + \mathbf{b}_o\right), \\
\mathbf{u}_t = & \tanh\left(\mathbf{W}_{ux}\mathbf{x}_t + \mathbf{W}_{uh}\mathbf{h}_{t-1} + \mathbf{b}_u\right), \\
\mathbf{c}_t = & \mathbf{i}_t \odot \mathbf{u}_t + \mathbf{f}_t \odot \mathbf{c}_{t-1}, \\
\mathbf{i}_t^* = & \sigma\left(\mathbf{W}_{ix}^*\mathbf{x}_t + \mathbf{W}_{ih}^*\mathbf{h}^* + \mathbf{b}_i^*\right), \\
\mathbf{f}_t^* = & \sigma\left(\mathbf{W}_{fx}^*\mathbf{x}_t + \mathbf{W}_{fh}^*\mathbf{h}^* + \mathbf{b}_f^*\right), \\
\mathbf{o}_t^* = & \sigma\left(\mathbf{W}_{ox}^*\mathbf{x}_t + \mathbf{W}_{oh}^*\mathbf{h}^* + \mathbf{b}_o^*\right), \\
\mathbf{u}_t^* = & \tanh\left(\mathbf{W}_{ux}^*\mathbf{x}_t + \mathbf{W}_{uh}^*\mathbf{h}^* + \mathbf{b}_u^*\right), \\
\mathbf{c}_t^* = & \mathbf{i}_t^* \odot \mathbf{u}_t^* + \mathbf{f}_t^* \odot \mathbf{c}^*, \\
\mathbf{h}_t = & \mathbf{o}_t \odot \tanh(\mathbf{c}_t) + \mathbf{o}_t^* \odot \tanh\left(\mathbf{c}_t^*\right),
\end{cases}
\tag{6.2}
$$

where $\mathbf{x}_t$ is the micro-video embedding at the time step t, $\mathbf{h}_{t-1}$ and $\mathbf{c}_{t-1}$ are, respectively, the hidden state and memory cell at the time step $t-1$, linking by edge $< v_c^{t_c-1}, v_c^{t_c} >$, and $\mathbf{h}^*$ and $\mathbf{c}^*$ are the hidden state and memory cell at the time step t^*, linking by edge $< v_c^{t_c^*}, v_c^{t_c} >$. Therefore, our temporal graph-based LSTM network can simultaneously leverage user's neighbor and cross-time interested context information to enhance the memorization of diverse interest and further strengthen the interest representation. And we can obtain the user's interested feature sequence $\mathbf{F}_{in} = [\mathbf{h}_{in,1}, \mathbf{h}_{in,2}, \dots, \mathbf{h}_{in,m_c}] \in \mathbb{R}^{d_c \times m_c}$, where d_c is the dimension of each hidden state in $\mathbf{F}_{in}$.

As the user's uninterested points are also dynamic and diverse, we build another temporal graph-based LSTM layer to model the user's $\mathcal{U}_n$ sequence and then obtain the uninterested feature sequence of the user, i.e., $\mathbf{F}_{un} = [\mathbf{h}_{un,1}, \mathbf{h}_{un,2}, \dots, \mathbf{h}_{un,m_n}] \in \mathbb{R}^{d_n \times m_n}$, where d_n is the dimension of each hidden state in $\mathbf{F}_{un}$.

6.4.2 THE MULTI-LEVEL INTEREST MODELING LAYER

Since there are multiple interactions between a user and a micro-video and they reflect different degrees of user's interest, we propose a multi-level interest modeling layer to further obtain the enhanced interest representation. As the "like" and "follow" behaviors indicate users' stronger interest compared with the "click" one, we hence utilize the "like" and "follow" information to enhance the interest representation. Particularly, for the user u, we set the weighted sum of micro-video representations in $\mathcal{U}_l$ and $\mathcal{U}_f$ as the user's enhanced interest feature $\mathbf{f}_{en}$, formulated as

$$\mathbf{f}_{en} = w_l \sum_{t_l=1}^{m_l} \mathbf{x}_l^{t_l} + w_f \sum_{t_f=1}^{m_f} \mathbf{x}_f^{t_f}, \tag{6.3}$$

where $\mathbf{x}_l^{t_l}$ is the embedding of micro-video $v_l^{t_l}$ in $\mathcal{U}_l$, $\mathbf{x}_f^{t_f}$ is the embedding of micro-video $v_f^{t_f}$ in $\mathcal{U}_f$, w_l, and w_f are the hyper parameters controlling the weights between "like" and "follow."

With the enhanced interest representation $\mathbf{f}_{en}$, we can construct an embedding matrix $\mathbf{U} \in \mathbb{R}^{N \times D}$, i.e., user matrix, where N and D, respectively, denote the number of users and the dimension of the enhanced interest representations. As the user's "like" and "follow" information more precisely indicates the user's interest, we can obtain more accurate interest representations using the user matrix. The user matrix $\mathbf{U}$ will be updated in the training phrase. Moreover, for each user, we utilize embedding lookup strategy to search the user's enhanced interest representation from the matrix $\mathbf{U}$ during the training and testing phrase.

6.4.3 THE PREDICTION LAYER

Standing on the shoulder of the user's interested feature sequence $\mathbf{F}_{in}$, uninterested feature sequence $\mathbf{F}_{un}$, and enhanced interest representation $\mathbf{f}_{en}$, we place a prediction layer to get the click probability of the given micro-video v_{new}, as shown in Figure 6.4. Specifically, we first feed $\mathbf{F}_{in}$ and the embedding of the given micro-video $\mathbf{x}_{new}$ into a vanilla attention layer to obtain the

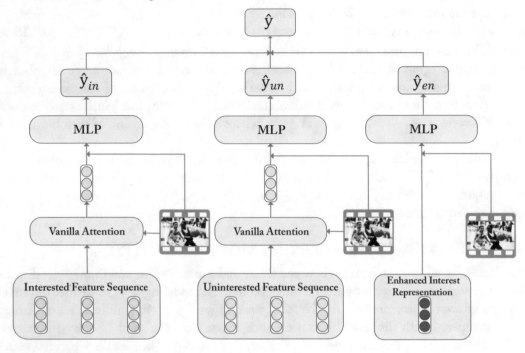

Figure 6.4: Structure of the Prediction Layer.

improved interested representation $\mathbf{f}_{in}$. Formally, the attention layer is defined as follows:

$$
\begin{cases}
\alpha_j = \dfrac{\exp\left(f\left(\mathbf{h}_{in,j}, \mathbf{x}_{new}\right)\right)}{\sum_{j=1}^{m_c} \exp\left(f\left(\mathbf{h}_{in,j}, \mathbf{x}_{new}\right)\right)}, \\
f\left(\mathbf{h}_{in,j}, \mathbf{x}_{new}\right) = \mathbf{h}_{in,j}^T \mathbf{W} \mathbf{x}_{new},
\end{cases}
\tag{6.4}
$$

where $\mathbf{h}_{in,j} \in \mathbb{R}^{d_c}$, $\mathbf{x}_{new} \in \mathbb{R}^D$, $\mathbf{W} \in \mathbb{R}^{d_c \times D}$, and α_j denotes the attention score of the jth interested feature in $\mathbf{F}_{in}$. With the attention weight α_j, the improved interested representation is computed as follows:

$$
\mathbf{f}_{in} = \sum_{j=1}^{m_c} \alpha_j \mathbf{h}_{in,j}.
\tag{6.5}
$$

Thereafter, we concatenate the improved interested representation $\mathbf{f}_{in}$ and the representation of the new micro-video x_{new}, and then feed it into a multi-layer perception (MLP) network, as follows:

$$
\begin{cases}
\mathbf{f}_1 = \phi\left(\mathbf{W}_1\left[\mathbf{f}_{in}, \mathbf{x}_{new}\right] + \mathbf{b}_1\right), \\
\hat{y}_{in} = \mathbf{W}_2 \mathbf{f}_1 + b_2,
\end{cases}
\tag{6.6}
$$

where $\mathbf{W}_1 \in \mathbb{R}^{d_c' \times (d_c + D)}$ and $\mathbf{W}_2 \in \mathbb{R}^{1 \times d_c'}$ denote the weight matrixes, $\mathbf{b}_1 \in \mathbb{R}^{d_c'}$ and b_2, respectively, denote the bias vector and the bias value, and ϕ denotes the ReLU activation function. $\hat{y}_{in}$ is the click probability calculated by the improved interested representation $\mathbf{f}_{in}$.

Similarly, we can obtain the improved uninterested representation $\mathbf{f}_{un}$ based on $\mathbf{F}_{un}$ and $\mathbf{x}_{new}$ using another vanilla attention layer. Afterward, we feed the concatenation of the improved uninterested representation $\mathbf{f}_{un}$ and the new micro-video embedding $\mathbf{x}_{new}$ into two MLP layers, and obtain the click probability based on the improved uninterested representation, i.e., $\hat{y}_{un}$. Analogously, the click probability based on the enhanced interest representation, i.e., $\hat{y}_{en}$, can be obtained by feeding the concatenation of the enhanced interest representation $\mathbf{f}_{en}$ and the new micro-video embedding $\mathbf{x}_{new}$ into two MLP layers.

Finally, the weighted sum of the above three probability values is set as our prediction result,

$$\hat{y} = \alpha_1 \hat{y}_{in} + \alpha_2 \hat{y}_{un} + \alpha_3 \hat{y}_{en}, \tag{6.7}$$

where α_1, α_2, and α_3 are the hyper parameters controlling the weights of $\hat{y}_{in}$, $\hat{y}_{un}$, and $\hat{y}_{en}$, respectively, and $\hat{y}$ is the final output of our model denoting the click probability of the given user on the given new micro-video.

Our method is trained as an end-to-end deep learning model equipped with the sigmoid cross-entropy loss:

$$L(\hat{y}) = -\left(y \log \sigma(\hat{y}) + (1 - y) \log(1 - \sigma(\hat{y})) \right), \tag{6.8}$$

where σ denotes the sigmoid activation function and $y \in \{0, 1\}$ is the ground truth that indicates whether the user clicks the micro-video or not. Besides, the back-propagation through time (BPTT) method is adopted to train our ALPINE model.

6.5   EXPERIMENTS

6.5.1   EXPERIMENTAL SETTINGS

Implementation Details. In the Dataset III-1, we utilized the 64-d visual embedding to represent the micro-video. As for the Dataset III-2, the concatenation of the 64-d category embedding and the 64-d visual embedding is set as the micro-video embedding. The length of users' historical sequence is set to 300. If it exceeds 300, we truncated it to 300; otherwise, we padded it to 300 and masked the padding in the network. We optimized the parameters using Adam with the initial learning rate 0.001, and the batch size is 2048.

6.5.2   BASELINES

To demonstrate the effectiveness of our proposed ALPINE model, we compared it with the following state-of-the-art methods.

- **BPR** [137]: This is a Bayesian personalized ranking model, which trains on pairwise items by maximizing the difference between the posterior probability of the positive samples and the negative ones.

- **CNN-R**: This model is a CNN-based recommendation system, which utilizes the CNN structure to model sequential information. In particular, it first applies different convolutional kernels to the sequential feature matrix. Explicitly, the window size varies from one to ten, and each kernel size has 32 linear filters. Thereafter, it feeds the obtained feature map into the max pooling layer followed by a fully connected layer to obtain interest embedding. Finally, a MLP is followed to predict the click probability.

- **LSTM-R**: This model utilizes the LSTM network to model the user's sequential information. Having obtained the hidden states, it feeds them into a fully connected layer to generate the interest representation, and then a MLP module is adopted to predict the click probability.

- **ATRank** [203]: It is an attention-based user behavior modeling framework, which captures the user's behavior interactions in multiple semantic spaces by the self-attention mechanism.

- **NCF** [60]: It is a collaborative filtering-based deep recommendation model, which learns the user embedding and the item embedding with a shallow network (element-wise product between user and item) and a deep network (concatenation of the user and item embedding followed by several MLP layers).

- **THACIL** [28]: It is a self-attention-based method for the micro-video recommendation, which utilizes a multi-head self-attention layer to capture the long-term correlation within user behaviors and the item and category two-level attention layer to model the fine-grained profiling of the user interest.

It is worth mentioning that THACIL and ATRank utilize the same click probability prediction layer as our model. As to the other methods including CNN-R, LSTM-R, BPR, and NCF, we fed the interest representations and the embedding of the new micro-video into the MLP layer to predict the click probability.

6.5.3   OVERALL COMPARISON

We conducted an empirical study to investigate whether our proposed model can achieve better recommendation performance. The results of all methods on two datasets are summarized in Table 6.1. Several observations stand out.

- BPR performs worse than the other baselines since it overlooks the sequential characteristic of the users' interest information. It hence fails to exploit the user's dynamic interest, revealing the necessity of modeling the historical sequence.

Table 6.1: Performance comparison between our proposed model and several state-of-the-art baselines over Dataset III-1 and III-2. And statistical significance over AUC between ALPINE and the best baseline (i.e., THACIL) is determined by a t-test ($\triangle$ denotes p-value <0.01).

Methods	Dataset III-1				Dataset III-2			
	AUC	P@50	R@50	F@50	AUC	P@50	R@50	F@50
BPR	0.595	0.290	0.387	0.331	0.583	0.241	0.181	0.206
LRTM-R	0.713	0.316	0.420	0.360	0.641	0.277	0.205	0.236
CNN-R	0.719	0.312	0.413	0.356	0.650	0.287	0.214	0.245
ATRank	0.722	0.322	0.426	0.367	0.660	0.297	0.221	0.253
NCF	0.724	0.320	0.420	0.364	0.672	0.316	0.225	0.262
THACIL	0.727	0.325	0.429	0.369	0.684	0.324	0.234	0.269
ALPINE	**0.739**$^\triangle$	**0.331**	**0.436**	**0.376**	**0.713**$^\triangle$	0.300	**0.460**	**0.362**

- Sequential modeling methods, including LSTM-R, CNN-R, ATRank, and THACIL, surpass the BPR model. This verifies the effectiveness of sequence modeling. Moreover, the self-attention based models, i.e., ATRank and THACIL, outperform CNN-R and LSTM-R, especially the latter one. It reveals that simply utilizing the LSTM network is insufficient to capture the users' dynamic and diverse interest information from a very long sequence. The attention mechanism can implicitly reduce the memorization length by focusing on the key interest information, that is why ATRank and THACIL achieve better performance on two datasets.

- While NCF does not model the user's historical information as a sequence, it also achieves promising performance compared with the other baselines. Probably because setting a user embedding matrix and updating it in the training stage can improve the interest representation. Moreover, two operations, the element wise product and several MLPs, model the relationship between users and items better.

- ALPINE achieves the best performance, substantially surpassing all the baselines. Particularly, ALPINE presents consistent improvements over sequential models like ATRank and THACIL, verifying the importance of memorizing the prior interested information and employing the temporal graph-based LSTM network on enhancing the interest representation. In addition, our proposed ALPINE exceeds NCF, because NCF randomly initializes the user matrix rather than explores its multi-level interest information. This justifies the effectiveness of our proposed multi-level interest modeling module. Moreover, as ALPINE also characterizes the user's uninterested cues, which can further improve the recommendation performance.

In addition, we also conducted the significance test between our model and the most competitive baseline THACIL. We can see that the advantage of our model is statistically significant as p-value is 2.81×10^{-5} on the Dataset III-1 and 4.70×10^{-6} on the Dataset III-2.

To justify the robustness of our proposed model, we comparatively explored the performance of our model and the baselines by varying the number of returned items K. Figure 6.5 shows the results regarding the performance comparison on K.

- Jointly analyzing the performance of the models in Figures 6.5a and d, we found that increasing the number of returned items K degrades the precision value of the recommendation. But our model ALPINE outperforms others under the same experimental setting, especially on the Dataset III-1.

- The performance of all these methods over recall and F value rises fast as the number of returned items K linearly increases. Their curves then gradually ascend to a steady state. Our method ALPINE consistently and remarkably outputs a higher accuracy as compared to that of other methods, especially on the Dataset-III-2. This verifies the robustness of our model.

6.5.4 COMPONENT-WISE EVALUATION OF ALPINE

We studied the variants of our model to further investigate the effectiveness of the uninterested representation modeling, user-matrix, and temporal interest graph:

- **ALPINE_u**: We eliminated the uninterested representation modeling part from the model. Namely, we computed the final click probability by interested representation and multi-level interest representation.

- **ALPINE_m**: We eliminated the multi-level interest module. That is, the final click probability is computed by the user's interested and uninterested representation.

- **ALPINE_um**: We only utilized the user's interested sequence to predict the click probability, namely we eliminated both the uninterested representation modeling and the multi-level interest modeling layer.

- **ALPINE_umg**: We eliminated the graph information from the ALPINE_um model.

We compared these variants on the two datasets, and Table 6.2 summarizes the results regarding the component-wise comparison. By jointly analyzing Table 6.2, we gained the following insights:

- By jointly analyzing the performance of ALPINE_u on the two datasets, it can be seen that removing the uninterested representation modeling degrades the recommendation results. To be more specific, ALPINE_u has dropped by 0.2% on the Dataset III-1 and 1.1% on the Dataset III-2 in terms of AUC. This verifies the effectiveness of the uninterested representation modeling.

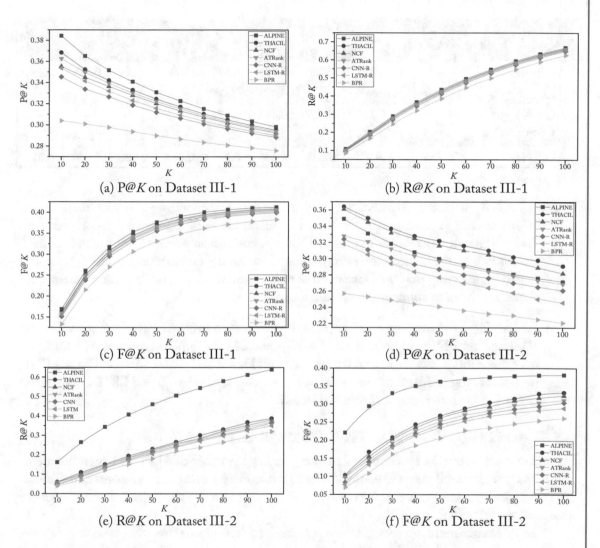

(a) P@K on Dataset III-1

(b) R@K on Dataset III-1

(c) F@K on Dataset III-1

(d) P@K on Dataset III-2

(e) R@K on Dataset III-2

(f) F@K on Dataset III-2

Figure 6.5: Recommendation performance vs. the number of returned items K over Datasets III-1 and III-2.

Table 6.2: Component-wise validation of our proposed ALPINE model over Datasets III-1 and III-2 by disabling one component each time. And statistical significance over AUC among all baselines is determined by a t-test ($\triangle$ denotes p-value <0.01 and $\diamond$ denotes p-value <0.05).

Methods	Dataset III-1				Dataset III-2			
	AUC	P@50	R@50	F@50	AUC	P@50	R@50	F@50
ALPINE_u	0.737$^\diamond$	0.330	0.435	0.375	0.702$^\triangle$	0.294	0.454	0.356
ALPINE_m	0.735$^\triangle$	0.329	0.433	0.374	–	–	–	–
ALPINE_um	0.734$^\triangle$	0.327	0.432	0.372	–	–	–	–
ALPINE_umg ALPINE_ug	0.716$^\triangle$	0.318	0.426	0.363	0.654$^\triangle$	0.291	0.219	0.250
ALPINE	**0.739**	**0.331**	**0.436**	**0.376**	**0.713**	**0.300**	**0.460**	**0.362**

- ALPINE surpasses ALPINE_m, indicating that incorporating the user matrix layer is beneficial to strengthen the interest representation. Moreover, compared with ALPINE_u, the performance of ALPINE_um conformably drops 0.3% under four metrics, which further reflects the effectiveness of our multi-level interest modeling layer. It is worth mentioning that the Dataset III-2 only contains "click" and "not click" interaction, therefore the corresponding results are vacant.

- ALPINE_um shows the consistent improvements over the ALPINE_umg on Dataset III-1 and ALPINE_ug on Dataset III-2. Specifically, the improvements of ALPINE_um over these models in terms of AUC are 2.3% on Dataset III-1 and 5.9% on Dataset III-2, demonstrating the great advantage of our novel temporal graph-based LSTM network on capturing both dynamic and diverse interest.

6.5.5 JUSTIFICATION OF THE TEMPORAL GRAPH

Apart from achieving the superior performance, the key advantage of ALPINE over other methods is that its temporal graph structure is able to strengthen the interest representation. Toward this end, we carried out experiments over the two datasets to verify the influence of the neighbor size L of the temporal graph.

In this experiment, we selected the top L similar micro-videos from the graph as neighbors of the given micro-video rather than considering the top one. Specifically, we set the average of the top L similar micro-videos' hidden states and memory cells as h^* and c^* in Eq. (6.2), respectively. The comparison results vs. the neighbor size L are illustrated in Figure 6.6. We found that the performance consistently drops under different evaluation metrics when L increases, especially the AUC drops significantly. This may be due to the fact that much more noise is introduced when a micro-video is connected with many others. Therefore, in this chapter, we set L equals to one.

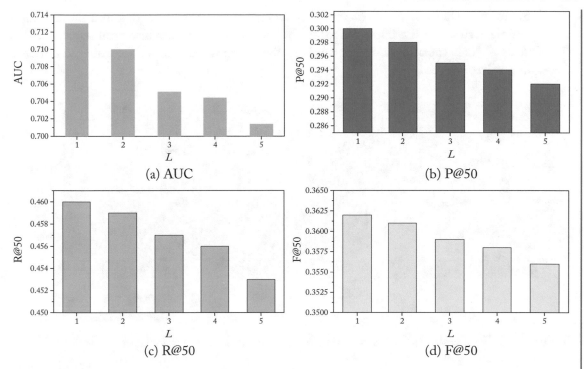

Figure 6.6: Illustration of the neighbor size L of the temporal graph on our recommendation performance.

6.5.6 ATTENTION VISUALIZATION

As analyzed before, we fed the interested feature sequence $\mathbf{F}_{in}$ and a new micro-video's embedding $\mathbf{x}_{new}$ into a vanilla attention layer to obtain the improved interested representation. To intuitively illustrate the attention results, we randomly selected some new micro-videos from the test data and visualized the attention scores in Figure 6.7. Several interesting observations stand out.

- For each new micro-video, the attention scores of its historical clicked micro-videos are different, which indicates that different micro-videos in the historical sequence contribute differently.

- By and large, the earlier a micro-video locates in the sequence, the smaller the attention score is, which indicates that the latter clicked micro-videos contribute more to the recommendation. This observation strongly supports that the user's interest is dynamic.

- By visualizing the categories of micro-videos, we noticed that, micro-videos from the same category contributes more to the recommendation results. As shown in the sub-figure of

Figure 6.7, the new given micro-video belongs to the 55th category, and the attention mainly focuses on the micro-videos of the same category in the historical sequence. This demonstrates the attention layer can help obtain improved features according to different new micro-videos.

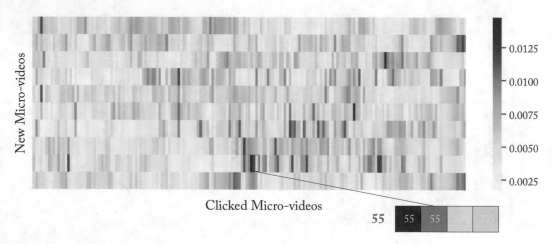

Figure 6.7: Visualization of the attention mechanism in the prediction layer.

6.6   SUMMARY

In this chapter, we present a temporal graph-based LSTM model to intelligently route micro-videos to the target users. To capture the users' dynamic and diverse interest, we encode their historical interaction sequence into a temporal graph and then design a novel temporal graph-based LSTM to model it. As different interactions reflect different degrees of interest, we build a multi-level interest modeling layer to enhance users' interest representation. Moreover, our model extracts uninterested information from true negative samples to improve the recommendation performance. To justify our scheme, we perform extensive experiments on two public datasets, and the experimental results demonstrate the effectiveness of our model.

CHAPTER 7

Research Frontiers

In this book, we investigate some application-motivated problems, namely the research problems of micro-video understanding. To solve these problems, we design some general principles, methodologies, and optimizations by jointly learning from multiple correlated modalities of the given micro-videos, including the textual, visual, acoustic, and social ones. They are empirically validated on multiple real-world datasets. In particular, we first introduce the proliferation of micro-video services and identify three practical tasks of micro-video understanding: popularity prediction, venue category estimation, and micro-video routing. Based upon these tasks, we analyze the unique research challenges of micro-videos that are distinct from traditional long videos, such as information sparseness, hierarchical structure, low-quality, multimodal sequential data, as well as lack of benchmark datasets. To address these problems, we present a series of multimodal learning methods, consisting of multimodal transductive learning, multimodal cooperative learning, multimodal transductive learning and multimodal sequential learning. These theoretical methods are verified over three datasets we constructed. To facilitate other researchers, we have released the codes, parameter settings, as well as the three datasets. We have to emphasize that learning from multiple modalities of the given micro-videos is still a young and highly promising research field. There are many unexplored but fruitful future directions and challenging research issues. We illustrate a few of them here.

7.1 MICRO-VIDEO ANNOTATION

Facing the exponentially growing number of micro-videos, it is important to help users quickly identify their desired ones. The hashtags associated with micro-videos are typically provided by uploaders to summarize the post content of users and attract the attention of followers. Taking the popular social platform Instagram as an example, as shown in Figure 7.1, the hashtags are prefixed with the symbol "#" to mark keywords or key topics of a post. The hashtags have been proved to be useful in many applications, including micro-blog retrieval, event analysis, and sentiment analysis. Moreover, the tagging service can benefit the stakeholders of micro-video ecosystems. For users, hashtags facilitate them to search and locate their desired micro-videos. For post-sharers, concise and concrete hashtags can increase the probability of their micro-videos to be discovered. For platforms, hashtags can make the management of micro-videos (e.g., categorization) more convenient. Despite their importance, numerous micro-videos are lack of hashtags or the hashtags are inaccurate or incomplete. In light of this, micro-video annotation,

Figure 7.1: An example of micro-video with hashtags in Instagram.

which suggests a list of hashtags to a user when he or she wants to annotate a post, becomes a crucial research problem.

Although several models have been adopted for hashtag recommendation and achieved some progress, such as collaborative filtering, generative models, and DNNs, they mainly focus on hashtag recommendation for micro-blogs or social images. Limited research efforts have been devoted to the micro-video annotation, due to the following reasons. (1) Long-tail distribution. The hashtag distribution is heavily skewed toward a few frequent hashtags with a long-tail consisting of less frequent tags, as shown in Figure 7.2. Current studies note that many hashtags from the long-tail are "misspelled" or "meaningless" words, we believe that there are some meaningful hashtags within the long-tail which have been overlooked. That is, how to create correlations between the frequent hashtags and their "related" long-tail hashtags to enhance the representation of them is untapped. (2) Multimodal sequence modeling. Micro-videos consist of visual, acoustic, and textual modalities, which are encoded together with sequential structure. On one hand, different streams in one micro-video demonstrate different temporal dynamics and thus should be modeled individually. For example, the objects in a micro-video could be the same throughout the time span of the micro-video, while the motion and audio may change from time to time. On the other hand, different modalities depict the intrinsic content of micro-videos consistently and complementarily from different views. Therefore, how to capture the sequential and multi-modality features is a considerable problem. (3) Diverse annotation. We find that

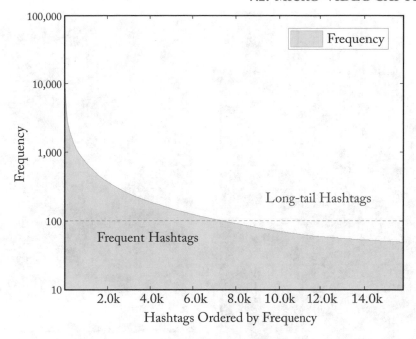

Figure 7.2: The statistic of hashtag frequency distribution in Instagram.

existing approaches recommend hashtags while ignoring redundancy among them, therefore, how to obtain relevant hashtags in consideration of their inter-dependencies is difficult.

In the future, we will tackle this task from the following three directions. First, we plan to construct a knowledge graph to explore hashtag correlations, and leverage existing structural knowledge to derive proper dependencies between frequent hashtags and long-tail hashtags. Second, we will introduce multi-level attention mechanism into the multimodal sequence model to focus on important cues among the sequential features and multi-modality features. Lastly, we expect to simulate how human annotators works and generate diverse and distinct micro-video annotation.

7.2   MICRO-VIDEO CAPTIONING

Micro-video captioning aims to auto generate textual descriptions for micro-videos. Some examples can be found in Figure 7.3. Due to its representation capability involving both computer vision and natural language processing techniques, the micro-video captioning shows great potential in aiding visually impaired people better understand visual contents. Moreover, it also plays a vital role in searching micro-videos and answering questions regarding micro-video contents. As users tend to submit queries and ask questions about micro-video clips through text-based keywords, a better content descriptor can promote the user satisfaction as well as loyalty for

Figure 7.3: Example of micro-video caption.

micro-video platforms. However, current micro-video systems (e.g., Vine, Instagram, Kuaishou, TikTok) lack of these content descriptions, resulting in performance degradation of micro-video retrieval and question-answering systems. Besides, some of the user-annotated captions are not adequate enough to correctly describe the micro-video contents. Therefore, it is crucial to develop micro-video captioning approaches to auto generate concise and accurate video descriptions.

Although micro-video captioning is an important research task in literature, there are some challenges:

(1) With the fast development of DNNs, employing more powerful network structures (e.g., graph neural networks, reinforcement learning techniques) to micro-video captioning will undoubtedly improve the model performance. (2) Normally the salient part inside a micro-video consists of a short video clips (e.g., 10 s), which fits well with the attention mechanism. Considering this, how to utilizing the attention mechanism to generate micro-video descriptions will be an important research problem. (3) Traditional video captioning is struggled with tedious description problem due to the limitation of training corpus. Therefore, the novel caption generation will be a potential direction for the micro-video captioning task. (4) Since construction of datasets is a fundamental problem in machine learning and current micro-video datasets are short of these captioning information, more abundant datasets will benefit further related

literatures and make a step forward of micro-video captioning tasks. (5) Different from image captioning, visual descriptions in micro-videos are relatively shorter and the number of ground truth descriptions is limited, which results in the infeasibility of traditional captioning evaluation metrics, e.g., BLEU, ROUGE, METEOR, and CIDEr. Therefore, developing new evaluation metrics fitting micro-video captioning should be a popular future topic.

7.3 MICRO-VIDEO THUMBNAIL SELECTION

To retain users' stickiness, beyond improving the quality of micro-videos, micro-video platforms and publishers have to draw users' eyes quickly [63]. As the most representative snapshot, the thumbnail summaries a micro-video visually and provides the first impression to the users, as shown in Figure 7.4. Moreover, studies report that the thumbnail is a crucial deciding factor in determining to watch a video or skip to another [13]. It means that an appealing thumbnail makes the micro-video more attractive. However, due to the inconvenient operation on smartphones or lack of experience, selecting a good thumbnail poses a challenge to users. Therefore, we suggest that an automatic thumbnail selection strategy is necessary to the micro-video sharing platform.

Figure 7.4: Exemplar demonstration of the micro-video thumbnail.

Although several pioneer efforts [53, 76, 93, 104, 194] have been dedicated to jointly consider the quality and representativeness for selecting the thumbnail, they ignored the fact that the thumbnail should reflect the publisher's preference and meet more users' interests. Considering such fact, it brings the following challenges to the task: (1) how to measure the publisher's preferences on the different frames extracted from the micro-video; (2) how to calculate the popularity of each frame according to the distribution of users' interests on the platform; and (3) to our knowledge, there is no such a suitable dataset to explore the micro-video thumbnail selection. Toward these challenges, amounts of micro-videos associated with the side-information (e.g., comments, publishers' profiles) are first collected to build a large-scale micro-video dataset for the thumbnail selection. Based on the constructed dataset, the publisher's preferences on different frames are captured by analyzing the multimodal content information of micro-video and the publisher's profile. Then, in order to calculate the popularity of each frame, the distribution of users' interests on the platform can be modeled via exploring the comments of published micro-videos. Ultimately, an attractive thumbnail can be obtained by combining the publisher's preference and popularity with the quality and representativeness.

7.4   SEMANTIC ONTOLOGY CONSTRUCTION

With the flourish of plentiful micro-video communities and exponentially growing number of micro-videos, it is necessary to help users quickly find the micro-videos they are interested. As is known to all, the classification to the labels of micro-video is an important aspect to save the user overwhelmed in large amounts of micro-videos. However, the labels of micro-videos are usually uploaded by users, whose laziness caused the existing micro-video only has the automatically acquired geographical label, but lack that of the semantic ontology, which plays a pivotal rule in the content-based micro-video search. In fact, labels of an instance is generally correlated with certain structure, such as the category-level hierarchical structure between cooking methods and cooking dishes, and so is the semantic ontology labels of the micro-video. Actually, the structured hierarchical labels are able to exclude several noise when searching the desired micro-videos. For example, in Figure 7.5, the cooking videos of same cooking method are gathered in one branch, and all the cooking videos are clustered in a tree, which is convenient for the user who is seeking for the micro-video about the cooking method instead of the review or exhibition of the food. In light of this, the hierarchical semantic ontology construction and classification based on it of the micro-video get our attention.

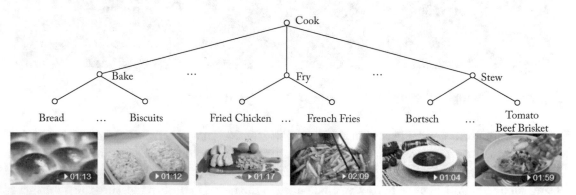

Figure 7.5: Example of hierarchical cooking videos.

Although several methods have been dedicated to the onefold semantic ontology classification of micro-video, the hierarchical semantic ontology classification remains largely untapped. Limited efforts have been devoted to the hierarchical semantic ontology classification of the micro-video, due to the following reasons: (1) the semantic ontology of the micro-video is massive, and the hierarchical structure is complicated, making it a great challenge to automatically construct a hierarchical ontology adapting to the micro-video classification; (2) the micro-video is composed of multiple modalities, such as visual, acoustic, and textual modalities. How to adaptively confuse the modalities to the multi-level hierarchical semantic ontology classification poses another challenge for us; and (3) although there are many public datasets toward the semantic ontology classification of micro-videos, there is a lack of the large-scale benchmark dataset for hierarchical ontology classification. Accordingly, how to construct a large-scale benchmark dataset to facilitate the implement and evaluation of the proposed research problem constitutes a tough challenge.

Toward this end, we will fulfill the aforementioned challenges from the following three directions: (1) we plan to construct the structured hierarchical ontology from the existing taxonomy knowledge adapted to the micro-video; (2) we will propose a modality-based and hierarchy-based attention mechanism into the hierarchical semantic ontology classification of the micro-video; and (3) we plan to construct a large-scale dataset from the online micro-video communities to facilitate studying the proposed research problem.

7.5   PORNOGRAPHIC CONTENT IDENTIFICATION

Though the exponentially growing micro-videos have brought prosperity to the new industry of micro-video platforms, they have brought the issue that anchorwomen propagate pornographic content through these platforms in order to attract people's eyeballs and earn more money. Pornographic content disrupts the cleanliness of the Internet, as juveniles are easy to expose to these unwholesome content and such phenomenon will pose a threat to the their physical and mental health. However, different from traditional porn videos, the anchorwomen on the micro-video platforms usually try to fool the platform by teasing the audience with their voices while keep their activities normal. Namely, traditional methods for filtering out inappropriate materials are inadequate for the micro-video platforms. In a sense, making further research on automatically detecting and isolating porn content for micro-video platforms is of immediate significance.

However, identifying pornographic videos from various micro-videos is non-trivial due to the following challenges. (1) For the micro-videos, identifying the voices of these anchorwomen is a pivotal issue. However, due to the complexity of social platforms, the cocktail effect of the complicated voice environments poses a challenge for automatic identification. To be specific, humans have the capacity to recognize what the other people are saying in a noisy crowd at a cocktail party as they can automatically exclude the unrelated sounds, while for machine, it is hard to select the related or unrelated information. (2) The increasing social slangs increase the difficulty of the identification, such as abbreviation and Internet buzzwords. For example, XSWL points to laughing my ass off while Europe represents the one who gets what he wants, which is even challenging for humans. (3) It is hard to distinguish the sexy voices and the normal voices only due to the audios. The modalities of the micro-videos consist of video, text, and audio. Therefore, how to jointly utilize different modalities for analyzing is a considerable problem.

In the future, we plan to tackle this from the following three aspects. First, we plan to simulate cocktail parties, namely, employing the mixture of artificial noises and the audios of the videos. Based on this, we will model the identification for individual voices and assignment for the voice track for each source. Second, we will introduce the voiceprint recognition for the selected audios to transforming the audios into text descriptions and filter out these audios without any semantics. Based on the above efforts, we plan to model the text anti-spam for text descriptions to identify the pornographic words and pure voice classification for these audios without semantics to decide whether erotic or not. Third, we will introduce multi-layer attention mechanisms into cross-modal learning of video, audio, and text for adaptively recognizing the sensitive and pornographic content of the micro-videos.

Bibliography

[1] S. Akaho. A kernel method for canonical correlation analysis. *IMPS*, 40(2):263–269, 2006. 15, 26, 93

[2] K. Ashraf, B. Elizalde, F. Iandola, M. Moskewicz, J. Bernd, G. Friedland, and K. Keutzer. Audio-based multimedia event detection with DNNs and sparse sampling. In *ACM ICMR*, pages 611–614, 2015. DOI: 10.1145/2671188.2749396 110

[3] F. R. Bach. Consistency of the group lasso and multiple kernel learning. In *JMLR*, vol. 9, pages 1179–1225, 2008. 65, 77

[4] F. R. Bach, G. R. G. Lanckriet, and M. I. Jordan. Multiple kernel learning, conic duality, and the SMO algorithm. In *ICML*, pages 6–13, 2004. DOI: 10.1145/1015330.1015424 16, 26

[5] Y. Bae and H. Lee. Sentiment analysis of twitter audiences: Measuring the positive or negative influence of popular twitterers. *Journal of the American Society for Information Science and Technology*, 63(12):2521–2535, 2012. DOI: 10.1002/asi.22768 24

[6] S. Bahrampour, N. M. Nasrabadi, A. Ray, and W. K. Jenkins. Multimodal task-driven dictionary learning for image classification. *IEEE Transactions on Image Processing*, 25(1):24–38, 2016. DOI: 10.1109/tip.2015.2496275 62, 82, 118

[7] S. Baluja, R. Seth, D. Sivakumar, Y. Jing, J. Yagnik, S. Kumar, D. Ravichandran, and M. Aly. Video suggestion and discovery for youtube: Taking random walks through the view graph. In *Proc. of the ACM International Conference on World Wide Web*, pages 895–904, 2008. DOI: 10.1145/1367497.1367618 125, 126

[8] J. Berger. Arousal increases social transmission of information. *Psychological Science*, 22(7):891–893, 2011. DOI: 10.1177/0956797611413294 23

[9] J. Berger and K. L. Milkman. What makes online content viral? *Journal of Marketing Research*, 49(2):192–205, 2012. DOI: 10.2139/ssrn.1528077 23

[10] S. Bhattacharya, B. Nojavanasghari, T. Chen, D. Liu, S.-F. Chang, and M. Shah. Towards a comprehensive computational model foraesthetic assessment of videos. In *Proc. of the ACM Multimedia Conference*, pages 361–364, 2013. DOI: 10.1145/2502081.2508119 22

[11] D. Borth, R. Ji, T. Chen, T. M. Breuel, and S. Chang. Large-scale visual sentiment ontology and detectors using adjective noun pairs. In *Proc. of the ACM Multimedia Conference*, pages 223–232, 2013. DOI: 10.1145/2502081.2502282 22

[12] S. Burger, Q. Jin, P. F. Schulam, and F. Metze. Noisemes: Manual annotation of environmental noise in audio streams. *Technical Report CMU-LTI-12–07*, pages 1–5, 2012. 15, 111

[13] G. Buscher, E. Cutrell, and M. R. Morris. What do you see when you're surfing?: Using eye tracking to predict salient regions of web pages. In *Proc. of the SIGCHI Conference on Human Factors in Computing Systems*, pages 21–30, ACM, 2009. DOI: 10.1145/1518701.1518705 145

[14] J.-F. Cai, E. J. Candès, and Z. Shen. A singular value thresholding algorithm for matrix completion. *SIAM Journal on Optimization*, 20(4):1956–1982, 2010. DOI: 10.1137/080738970 45

[15] S. Cao and N. Snavely. Graph-based discriminative learning for location recognition. In *IEEE CVPR*, pages 700–707, 2013. DOI: 10.1109/cvpr.2013.96 60, 61

[16] S. Cappallo, T. Mensink, and C. G. Snoek. Latent factors of visual popularity prediction. In *Proc. of International Conference on Multimedia Retrieval*, pages 195–202, 2015. DOI: 10.1145/2671188.2749405 24

[17] D. Castan and M. Akbacak. Segmental-GMM approach based on acoustic concept segmentation. In *SLAM@ INTERSPEECH*, pages 15–19, 2013. 110

[18] M. Cha, H. Kwak, P. Rodriguez, Y.-Y. Ahn, and S. Moon. I tube, you tube, everybody tubes: Analyzing the world's largest user generated content video system. In *Proc. of ACM SIGCOMM Conference on Internet Measurement*, pages 1–14, 2007. DOI: 10.1145/1298306.1298309 24

[19] K. Chaudhuri, S. M. Kakade, K. Livescu, and K. Sridharan. Multi-view clustering via canonical correlation analysis. In *Proc. of the International Conference on Machine Learning*, pages 129–136, ACM, 2009. DOI: 10.1145/1553374.1553391 26

[20] S. Chaudhuri and B. Raj. Unsupervised structure discovery for semantic analysis of audio. In *NIPS*, pages 1178–1186, 2012. 110

[21] B.-C. Chen, Y.-Y. Chen, F. Chen, and D. Joshi. Business-aware visual concept discovery from social media for multimodal business venue recognition. In *AAAI*, pages 61–68, 2016. 61

[22] C.-F. Chen, C.-P. Wei, and Y.-C. F. Wang. Low-rank matrix recovery with structural incoherence for robust face recognition. In *Proc. of IEEE Conference on Computer Vision and Pattern Recognition*, pages 2618–2625, 2012. DOI: 10.1109/cvpr.2012.6247981 26

[23] D. M. Chen, G. Baatz, K. Köser, S. S. Tsai, R. Vedantham, T. Pylvä, K. Roimela, X. Chen, J. Bach, M. Pollefeys, et al. City-scale landmark identification on mobile devices. In *IEEE CVPR*, pages 737–744, 2011. DOI: 10.1109/cvpr.2011.5995610 60, 61

[24] J. Chen, X. Song, L. Nie, X. Wang, H. Zhang, and T.-S. Chua. Micro tells macro: Predicting the popularity of micro-videos via a transductive model. In *Proc. of ACM International Conference on Multimedia*, pages 898–907, 2016. DOI: 10.1145/2964284.2964314 20, 24, 54

[25] J. Chen, H. Zhang, X. He, L. Nie, W. Liu, and T.-S. Chua. Attentive collaborative filtering: Multimedia recommendation with item-and component-level attention. In *Proc. of the International ACM SIGIR Conference on Research and Development in Information Retrieval*, pages 335–344, 2017. DOI: 10.1145/3077136.3080797 125, 126

[26] J. Chen, J. Zhou, and J. Ye. Integrating low-rank and group-sparse structures for robust multi-task learning. In *ACM KDD*, pages 42–50, 2011. DOI: 10.1145/2020408.2020423 69

[27] N. Chen, J. Zhu, and E. P. Xing. Predictive subspace learning for multi-view data: A large margin approach. In *NIPS*, pages 361–369, 2010. 61, 63

[28] X. Chen, D. Liu, Z.-J. Zha, W. Zhou, Z. Xiong, and Y. Li. Temporal hierarchical attention at category-and item-level for micro-video click-through prediction. In *Proc. of the ACM International Conference on Multimedia*, pages 1146–1153, 2018. DOI: 10.1145/3240508.3240617 16, 126, 127, 134

[29] J. Choi, G. Friedland, V. Ekambaram, and K. Ramchandran. Multimodal location estimation of consumer media: Dealing with sparse training data. In *IEEE ICME*, pages 43–48, 2012. DOI: 10.1109/icme.2012.141 61

[30] W. Chong, D. Blei, and F.-F. Li. Simultaneous image classification and annotation. In *IEEE Conference on Computer Vision and Pattern Recognition*, pages 1903–1910, 2009. DOI: 10.1109/cvprw.2009.5206800 74

[31] C. M. Christoudias, R. Urtasun, A. Kapoorz, and T. Darrell. Co-training with noisy perceptual observations. In *CVPR*, pages 2844–2851, 2016. DOI: 10.1109/cvpr.2009.5206572 25

[32] D. Coppersmith and S. Winograd. Matrix multiplication via arithmetic progressions. In *Proc. of ACM Symposium on Theory of Computing*, pages 1–6, 1987. DOI: 10.1016/s0747-7171(08)80013-2 56

[33] D. J. Crandall, L. Backstrom, D. Huttenlocher, and J. Kleinberg. Mapping the world's photos. In *ACM WWW*, pages 761–770, 2009. DOI: 10.1145/1526709.1526812 61

[34] P. Cui, Z. Wang, and Z. Su. What videos are similar with you?: Learning a common attributed representation for video recommendation. In *ACM MM*, pages 597–606, 2014. DOI: 10.1145/2647868.2654946 125, 126

[35] A. Culotta, N. K. Ravi, and J. Cutler. Predicting the demographics of twitter users from website traffic data. In *National Conference of the American Association for Artificial Intelligence*, pages 72–78, 2015. 82, 118

[36] S. Dhar, V. Ordonez, and T. L. Berg. High level describable attributes for predicting aesthetics and interestingness. In *Proc. of the IEEE Conference on Computer Vision and Pattern Recognition*, pages 1657–1664, 2011. DOI: 10.1109/cvpr.2011.5995467 22

[37] T. Diethe, D. R. Hardoon, and J. Shawe-Taylor. Multiview Fisher discriminant analysis. In *NIPS*, pages 1–8, 2008. 16, 26

[38] W. Ding, Y. Shang, L. Guo, X. Hu, R. Yan, and T. He. Video popularity prediction by sentiment propagation via implicit network. In *Proc. of ACM International on Conference on Information and Knowledge Management*, pages 1621–1630, 2015. DOI: 10.1145/2806416.2806505 24

[39] Z. Ding and Y. Fu. Low-rank common subspace for multi-view learning. In *Proc. of IEEE International Conference on Data Mining*, pages 110–119, 2014. DOI: 10.1109/icdm.2014.29 27

[40] Z. Ding, M. Shao, and Y. Fu. Latent low-rank transfer subspace learning for missing modality recognition. In *Proc. of AAAI Conference on Artificial Intelligence*, pages 1192–1198, 2014. 26

[41] Z. Ding and Y. Fu. Robust multi-view subspace learning through dual low-rank decompositions. In *Proc. of AAAI Conference on Artificial Intelligence*, pages 1181–1187, 2016. 27

[42] S. K. D'Mello and J. Kory. A review and meta-analysis of multimodal affect detection systems. *ACM Computing Surveys*, 47(3):1–36, 2015. DOI: 10.1145/2682899 25

[43] J. Donahue, Y. Jia, O. Vinyals, J. Hoffman, N. Zhang, E. Tzeng, and T. Darrell. Decaf: A deep convolutional activation feature for generic visual recognition. In *ICML*, pages 647–655, 2014. 118

[44] C. Dong, C. C. Loy, K. He, and X. Tang. Image super-resolution using deep convolutional networks. *IEEE Transactions on Pattern Analysis and Machine Intelligence*, 38(2):295–307, 2016. DOI: 10.1109/tpami.2015.2439281 85

[45] L. Duan, D. Xu, I. W. Tsang, and J. Luo. Visual event recognition in videos by learning from web data. *TPAMI*, 34(9):1667–1680, 2012. DOI: 10.1109/cvpr.2010.5539870 26

[46] M. Elad and M. Aharon. Image denoising via sparse and redundant representations over learned dictionaries. *IEEE Transactions on Image Processing*, 15(12):3736–3745, 2006. DOI: 10.1109/tip.2006.881969 62

[47] F. Feng, L. Nie, X. Wang, R. Hong, and T. S. Chua. Computational social indicators: A case study of Chinese university ranking. In *SIGIR*, pages 455–464, 2017. DOI: 10.1145/3077136.3080773 92

[48] A. Ferracani, D. Pezzatini, M. Bertini, and A. Del Bimbo. Item-based video recommendation: An hybrid approach considering human factors. In *Proc. of the ACM on International Conference on Multimedia Retrieval*, pages 351–354, 2016. DOI: 10.1145/2911996.2912066 15, 125, 126
DOI: 10.1145/2072609.2072619

[49] G. Friedland, J. Choi, H. Lei, and A. Janin. Multimodal location estimation on flickr videos. In *Proc. of the ACM SIGMM International Workshop on Social Media*, pages 23–28, 2011. DOI: 10.1145/2072609.2072619 4, 61

[50] G. Friedland, O. Vinyals, and T. Darrell. Multimodal location estimation. In *ACM MM*, pages 1245–1252, 2010. DOI: 10.1145/1873951.1874197 61

[51] H. Gao, F. Nie, X. Li, and H. Huang. Multi-view subspace clustering. In *IEEE International Conference on Computer Vision*, pages 4238–4246, 2015. DOI: 10.1109/iccv.2015.482 26

[52] J. Gao, T. Zhang, and C. Xu. A unified personalized video recommendation via dynamic recurrent neural networks. In *Proc. of the ACM International Conference on Multimedia*, pages 127–135, 2017. DOI: 10.1145/3123266.3123433 126, 127

[53] Y. Gao, T. Zhang, and J. Xiao. Thematic video thumbnail selection. In *16th IEEE International Conference on Image Processing (ICIP)*, pages 4333–4336, 2009. DOI: 10.1109/icip.2009.5419128 145
DOI: 10.1145/2733373.2806361

[54] F. Gelli, T. Uricchio, M. Bertini, A. Del Bimbo, and S.-F. Chang. Image popularity prediction in social media using sentiment and context features. In *Proc. of ACM International Conference on Multimedia*, pages 907–910, 2015. DOI: 10.1145/2733373.2806361 22, 24, 42

[55] S. Gopal and Y. Yang. Recursive regularization for large-scale classification with hierarchical and graphical dependencies. In *ACM KDD*, pages 257–265, 2013. DOI: 10.1145/2487575.2487644 68

[56] Y. Guo. Convex subspace representation learning from multi-view data. In *AAAI*, pages 2–9, 2013. 63

[57] L. Han and Y. Zhang. Learning tree structure in multi-task learning. In *ACM KDD*, pages 397–406, 2015. DOI: 10.1145/2783258.2783393 67

[58] J. Hays and A. A. Efros. Im2gps: Estimating geographic information from a single image. In *IEEE CVPR*, pages 1–8, 2008. DOI: 10.1109/cvpr.2008.4587784 60, 61

[59] J. He and R. Lawrence. A graph-based framework for multi-task multi-view learning. In *ICML*, pages 25–32, 2011. 61

[60] X. He, L. Liao, H. Zhang, L. Nie, X. Hu, and T. S. Chua. Neural collaborative filtering. In *ICLR*, pages 173–182, 2017. DOI: 10.1145/3038912.3052569 126, 134

[61] X. He, H. Zhang, M.-Y. Kan, and T.-S. Chua. Fast matrix factorization for online recommendation with implicit feedback. In *Proc. of the International ACM SIGIR Conference on Research and Development in Information Retrieval*, pages 549–558, 2016. DOI: 10.1145/2911451.2911489 126

[62] S. Hershey, S. Chaudhuri, D. P. W. Ellis, J. F. Gemmeke, A. Jansen, R. C. Moore, M. Plakal, D. Platt, R. A. Saurous, and B. Seybold. CNN architectures for large-scale audio classification. In *ICASSP*, pages 131–135, 2017. DOI: 10.1109/icassp.2017.7952132 105

[63] W. Hoiles, A. Aprem, and V. Krishnamurthy. Engagement and popularity dynamics of youtube videos and sensitivity to meta-data. *IEEE Transactions on Knowledge and Data Engineering*, 29(7):1426–1437, 2017. DOI: 10.1109/tkde.2017.2682858 145

[64] C. Hori, T. Hori, T.-Y. Lee, Z. Zhang, B. Harsham, J. R. Hershey, T. K. Marks, and K. Sumi. Attention-based multimodal fusion for video description. In *ICCV*, pages 4203–4212, IEEE, 2017. DOI: 10.1109/iccv.2017.450 99

[65] H. Hotelling. Relations between two sets of variates. *Biometrika*, 28(3/4):321–377, 1936. DOI: 10.2307/2333955 26, 39

[66] Y. Hu, Y. Koren, and C. Volinsky. Collaborative filtering for implicit feedback datasets. In *Proc. of the IEEE International Conference on Data Mining*, pages 263–272, 2008. DOI: 10.1109/icdm.2008.22 126

[67] G.-B. Huang, H. Zhou, X. Ding, and R. Zhang. Extreme learning machine for regression and multiclass classification. *IEEE Transactions on Systems, Man, and Cybernetics, Part B (Cybernetics)*, 42(2):513–529, 2012. DOI: 10.1109/tsmcb.2011.2168604 54, 55, 56

[68] G.-B. Huang, Q.-Y. Zhu, and C.-K. Siew. Extreme learning machine: A new learning scheme of feedforward neural networks. In *Proc. IEEE International Joint Conference on Neural Networks*, vol. 2, pages 985–990, 2004. DOI: 10.1109/ijcnn.2004.1380068 55

[69] Y. Huang, B. Cui, J. Jiang, K. Hong, W. Zhang, and Y. Xie. Real-time video recommendation exploration. In *Proc. of the ACM International Conference on Management of Data*, pages 35–46, 2016. DOI: 10.1145/2882903.2903743 125, 126

[70] L. Jacob, J.-p. Vert, and F. R. Bach. Clustered multi-task learning: A convex formulation. In *NIPS*, pages 745–752, 2009. 64

[71] Y. Jia, E. Shelhamer, J. Donahue, S. Karayev, J. Long, R. B. Girshick, S. Guadarrama, and T. Darrell. Caffe: Convolutional architecture for fast feature embedding. In *Proc. of the ACM Multimedia Conference*, pages 675–678, 2014. DOI: 10.1145/2647868.2654889 22, 33

[72] X. Jin, F. Zhuang, S. Wang, Q. He, and Z. Shi. Shared structure learning for multiple tasks with multiple views. In *MLKDD*, pages 353–368, 2013. DOI: 10.1007/978-3-642-40991-2_23 62

[73] X.-Y. Jing, R. Hu, F. Wu, X.-L. Chen, Q. Liu, and Y.-F. Yao. Uncorrelated multi-view discrimination dictionary learning for recognition. In *National Conference of the American Association for Artificial Intelligence*, pages 2787–2795, 2014. 62

[74] I. Jolliffe. *Principal Component Analysis*. Wiley Online Library, 2002. DOI: 10.1007/978-3-642-04898-2_455 26
DOI: 10.1109/TPAMI.2015.2435740

[75] M. Kan, S. Shan, H. Zhang, S. Lao, and X. Chen. Multi-view discriminant analysis. *IEEE Transactions on Pattern Analysis and Machine Intelligence*, 38(1):188–194, 2016. DOI: 10.1007/978-3-642-33718-5_58 34, 54, 55, 69

[76] H.-W. Kang and X.-S. Hua. To learn representativeness of video frames. In *Proc. of the 13th Annual ACM International Conference on Multimedia*, pages 423–426, 2005. DOI: 10.1145/1101149.1101242 145

[77] A. Khosla, A. Das Sarma, and R. Hamid. What makes an image popular? In *Proc. of ACM International Conference on World Wide Web*, pages 867–876, 2014. DOI: 10.1145/2566486.2567996 21, 24
DOI: 10.1145/2566486.2567996.

[78] A. Kilgarriff and C. Fellbaum. Wordnet: An electronic lexical database, 2000. DOI: 10.2307/417141 111

[79] S. Kim and E. P. Xing. Tree-guided group lasso for multi-task regression with structured sparsity. In *ICML*, pages 1–8, 2010. 64

[80] D. P. Kingma and J. Ba. Adam: A method for stochastic optimization. *NIPS*, pages 1–1, 2015. 99

[81] S. Kong and D. Wang. A dictionary learning approach for classification separating the particularity and the commonality. In *European Conference on Computer Vision*, pages 186–199, 2012. DOI: 10.1007/978-3-642-33718-5_14 62

[82] A. Krizhevsky, I. Sutskever, and G. E. Hinton. Imagenet classification with deep convolutional neural networks. In *Proc. of the Annual Conference on Neural Information Processing Systems*, pages 1106–1114, NIPS Foundation, 2012. DOI: 10.1145/3065386 22, 111

[83] G. R. G. Lanckriet, N. Cristianini, P. Bartlett, L. E. Ghaoui, and M. I. Jordan. Learning the kernel matrix with semidefinite programming. *JMLR*, 5(1):27–72, 2002. 26

[84] H. Lei, J. Choi, and G. Friedland. Multimodal city-verification on flickr videos using acoustic and textual features. In *IEEE ICASSP*, pages 2273–2276, 2012. DOI: 10.1109/icassp.2012.6288367 61

[85] H. Li, X. Ma, F. Wang, J. Liu, and K. Xu. On popularity prediction of videos shared in online social networks. In *Proc. of ACM International Conference on Information and Knowledge Management*, pages 169–178, 2013. DOI: 10.1145/2505515.2505523 24

[86] J. Li, Y. Wu, J. Zhao, and K. Lu. Low-rank discriminant embedding for multiview learning. *IEEE Transactions on Cybernetics*, pages 1–14, 2016. DOI: 10.1109/tcyb.2016.2565898 27

[87] S. Li and Y. Fu. Learning balanced and unbalanced graphs via low-rank coding. *IEEE Transactions on Knowledge and Data Engineering*, 27(5):1274–1287, 2015. DOI: 10.1109/tkde.2014.2365793 39

[88] Z. Li, J. Wang, J. Cai, Z. Duan, H. Wang, and Y. Wang. Non-reference audio quality assessment for online live music recordings. In *Proc. of the ACM Multimedia Conference*, pages 63–72, 2013. DOI: 10.1145/2502081.2502106 21, 23

[89] T.-Y. Lin, Y. Cui, S. Belongie, and J. Hays. Learning deep representations for ground-to-aerial geolocalization. In *IEEE CVPR*, pages 5007–5015, 2015. DOI: 10.1109/cvpr.2015.7299135 61

[90] Z. Lin, M. Chen, and Y. Ma. The augmented lagrange multiplier method for exact recovery of corrupted low-rank matrices. *Technical Report, UILU-ENG-09–2215*, 2010. 45

[91] A. Liu, W. Nie, Y. Gao, and Y. Su. Multi-modal clique-graph matching for view-based 3d model retrieval. *IEEE Transactions on Image Processing*, 25(5):2103–2116, 2016. DOI: 10.1109/tip.2016.2540802 26

[92] A. Liu, Z. Wang, W. Nie, and Y. Su. Graph-based characteristic view set extraction and matching for 3D model retrieval. *Information Sciences*, 320:429–442, 2015. DOI: 10.1016/j.ins.2015.04.042 26

[93] C. Liu, Q. Huang, and S. Jiang. Query sensitive dynamic web video thumbnail generation. In *18th IEEE International Conference on Image Processing*, pages 2449–2452, 2011. DOI: 10.1109/icip.2011.6116155 145

[94] G. Liu, Z. Lin, S. Yan, J. Sun, Y. Yu, and Y. Ma. Robust recovery of subspace structures by low-rank representation. *IEEE Transactions on Pattern Analysis and Machine Intelligence*, 35(1):171–184, 2013. DOI: 10.1109/tpami.2012.88 56

[95] G. Liu, Z. Lin, and Y. Yu. Robust subspace segmentation by low-rank representation. In *Proc. of International Conference on Machine Learning*, pages 663–670, 2010. 26, 27, 39

[96] G. Liu and S. Yan. Latent low-rank representation for subspace segmentation and feature extraction. In *Proc. of IEEE International Conference on Computer Vision*, pages 1615–1622, 2011. DOI: 10.1109/iccv.2011.6126422 26, 27, 39

[97] G. Liu, Y. Yan, E. Ricci, Y. Yang, Y. Han, S. Winkler, and N. Sebe. Inferring painting style with multi-task dictionary learning. In *International Joint Conference on Artificial Intelligence*, pages 2162–2168, 2015. 62

[98] H. Liu, X. Yang, L. J. Latecki, and S. Yan. Dense neighborhoods on affinity graph. In *IJCV*, vol. 98, pages 65–82, 2012. DOI: 10.1007/s11263-011-0496-1 67

[99] J. Liu, S. Ji, and J. Ye. Multi-task feature learning via efficient l2, 1-norm minimization. In *UAI*, pages 339–348, 2009. 68, 82

[100] J. Liu, Y. Yang, Z. Huang, and Y. Yang. On the influence propagation of web videos. *IEEE Transactions on Knowledge and Data Engineering*, 26(8):1961–1973, 2014. DOI: 10.1109/tkde.2013.142 24

[101] M. Liu, Y. Luo, D. Tao, C. Xu, and Y. Wen. Low-rank multi-view learning in matrix completion for multi-label image classification. In *Proc. of AAAI Conference on Artificial Intelligence*, pages 2778–2784, 2015. 27

158 BIBLIOGRAPHY

[102] M. Liu, L. Nie, M. Wang, and B. Chen. Towards micro-video understanding by joint sequential-sparse modeling. In *Proc. of the ACM International Conference on Multimedia*, pages 970–978, 2017. DOI: 10.1145/3123266.3123341 90

[103] M. Liu, X. Wang, L. Nie, X. He, B. Chen, and T.-S. Chua. Attentive moment retrieval in videos. In *Proc. of the ACM International Conference on Conference on Research and Development in Information Retrieval*, pages 15–24, 2018. DOI: 10.1145/3209978.3210003 59

[104] W. Liu, T. Mei, Y. Zhang, C. Che, and J. Luo. Multi-task deep visual-semantic embedding for video thumbnail selection. In *Proc. of the IEEE Conference on Computer Vision and Pattern Recognition*, pages 3707–3715, 2015. DOI: 10.1109/cvpr.2015.7298994 145

[105] M. Long, Y. Cao, J. Wang, and M. I. Jordan. Learning transferable features with deep adaptation networks. In *ICML*, pages 97–105, 2015. 118

[106] Y. Lu, Z. Lai, Y. Xu, X. Li, D. Zhang, and C. Yuan. Low-rank preserving projections. *IEEE Transactions on Cybernetics*, 46(8):1900–1913, 2016. DOI: 10.1109/tcyb.2015.2457611 26, 39

[107] Z. Ma, A. Sun, and G. Cong. On predicting the popularity of newly emerging hashtags in twitter. *Journal of the American Society for Information Science and Technology*, 64(7):1399–1410, 2013. DOI: 10.1002/asi.22844 24

[108] J. Mairal and F. Bach. Task-driven dictionary learning. *IEEE Transactions on Pattern Analysis and Machine Intelligence*, 34(4):791–804, 2012. DOI: 10.1109/tpami.2011.156 62

[109] J. Mairal, F. Bach, J. Ponce, and G. Sapiro. Online dictionary learning for sparse coding. In *Proc. of the Annual International Conference on Machine Learning*, pages 689–696, 2009. DOI: 10.1145/1553374.1553463 62, 79, 81, 82

[110] J. Mairal, M. Elad, and G. Sapiro. Sparse representation for color image restoration. *IEEE Transactions on Image Processing*, 17(1):53–69, 2008. DOI: 10.1109/TIP.2007.911828 62

[111] J. Mairal, J. Ponce, G. Sapiro, A. Zisserman, and F. R. Bach. Supervised dictionary learning. In *Proc. of the Annual Advances in Neural Information Processing Systems*, pages 1033–1040, 2009. 62

[112] P. J. McParlane, Y. Moshfeghi, and J. M. Jose. Nobody comes here anymore, it's too crowded; predicting image popularity on flickr. In *Proc. of International Conference on Multimedia Retrieval*, pages 385–391, 2014. DOI: 10.1145/2578726.2578776 24

[113] T. Mei, B. Yang, X.-S. Hua, and S. Li. Contextual video recommendation by multimodal relevance and user feedback. *ACM Transactions on Information Systems*, 29(2):10, 2011. DOI: 10.1145/1961209.1961213 125, 126

[114] A. Mesaros, T. Heittola, A. Eronen, and T. Virtanen. Acoustic event detection in real life recordings. In *IEEE EUSIPCO*, pages 1267–1271, 2010. 111

[115] T. Mikolov, I. Sutskever, K. Chen, G. Corrado, and J. Dean. Distributed representations of words and phrases and their compositionality. In *Proc. of the Annual Conference on Neural Information Processing Systems*, pages 3111–3119, NIPS Foundation, 2013. 23, 115

[116] M.-F. Moens, K. Pastra, K. Saenko, and T. Tuytelaars. Vision and language integration meets multimedia fusion. In *Proc. of the ACM International Conference on Multimedia*, page 1493, 2016. DOI: 10.1145/2964284.2980537 59

[117] G. Monaci, P. Jost, P. Vandergheynst, B. Mailhe, S. Lesage, and R. Gribonval. Learning multimodal dictionaries. *IEEE TIP*, 16(9):2272–2283, 2007. DOI: 10.1109/tip.2007.901813 118

[118] E. Morvant, A. Habrard, S. Ayache, and phane. *Majority Vote of Diverse Classifiers for Late Fusion*. Springer, 2014. DOI: 10.1007/978-3-662-44415-3_16 25

[119] J. Ngiam, A. Khosla, M. Kim, J. Nam, H. Lee, and A. Y. Ng. Multimodal deep learning. In *ICML*, pages 689–696, 2011. 98

[120] P. X. Nguyen, G. Rogez, C. C. Fowlkes, and D. Ramanan. The open world of micro-videos. *CoRR*, abs/1603.09439, 2016. 3

[121] L. Nie, M. Wang, L. Zhang, S. Yan, B. Zhang, and T.-S. Chua. Disease inference from health-related questions via sparse deep learning. *IEEE Transactions on Knowledge and Data Engineering*, 27(8):2107–2119, 2015. DOI: 10.1109/tkde.2015.2399298 42

[122] L. Nie, X. Wang, J. Zhang, X. He, H. Zhang, R. Hong, and Q. Tian. Enhancing micro-video understanding by harnessing external sounds. In *ACM MM*, pages 1192–1200, 2017. DOI: 10.1145/3123266.3123313 90, 99, 101

[123] L. Nie, L. Zhang, Y. Yang, M. Wang, R. Hong, and T.-S. Chua. Beyond doctors: Future health prediction from multimedia and multimodal observations. In *Proc. of the ACM Multimedia Conference*, pages 591–600, 2015. DOI: 10.1145/2733373.2806217 30, 34, 83

[124] L. Nie, Y.-L. Zhao, M. Akbari, J. Shen, and T.-S. Chua. Bridging the vocabulary gap between health seekers and healthcare knowledge. *IEEE Transactions on Knowledge and Data Engineering*, 27(2):396–409, 2015. DOI: 10.1109/tkde.2014.2330813 51

[125] K. Nigam and R. Ghani. Analyzing the effectiveness and applicability of co-training. In *CIKM*, pages 86–93, 2000. DOI: 10.1145/354756.354805 25

[126] Q. Pan, D. Kong, C. H. Ding, and B. Luo. Robust non-negative dictionary learning. In *National Conference of the American Association for Artificial Intelligence*, pages 2027–2033, 2014. 62

[127] R. Pan, Y. Zhou, B. Cao, N. N. Liu, R. Lukose, M. Scholz, and Q. Yang. One-class collaborative filtering. In *Proc. of the IEEE International Conference on Data Mining*, pages 502–511, 2008. DOI: 10.1109/icdm.2008.16 126

[128] S. L. Pancoast, M. Akbacak, and M. H. Sanchez. Supervised acoustic concept extraction for multimedia event detection. In *Proc. of the ACM International Workshop on Audio and Multimedia Methods for Large-Scale Video Analysis*, pages 9–14, 2012. DOI: 10.1145/2390214.2390219 110, 111

[129] J. Park, S.-J. Lee, S.-J. Lee, K. Kim, B.-S. Chung, and Y.-K. Lee. Online video recommendation through tag-cloud aggregation. *IEEE Transaction on MultiMedia*, 18(1):78–87, 2011. DOI: 10.1109/mmul.2010.6 125, 126

[130] F. Pedregosa, G. Varoquaux, A. Gramfort, V. Michel, B. Thirion, O. Grisel, M. Blondel, P. Prettenhofer, R. Weiss, V. Dubourg, et al. Scikit-learn: Machine learning in Python. In *Journal of Machine Learning Research, (JMLR)*, vol. 12, pages 2825–2830, 2011. 34, 68

[131] H. Peng, K. Li, B. Li, H. Ling, W. Xiong, and W. Hu. Predicting image memorability by multi-view adaptive regression. In *Proc. ACM International Conference on Multimedia*, pages 1147–1150, 2015. DOI: 10.1145/2733373.2806303 42

[132] M. Quadrana, A. Karatzoglou, B. Hidasi, and P. Cremonesi. Personalizing session-based recommendations with hierarchical recurrent neural networks. In *Proc. of the 11th ACM Conference on Recommender Systems*, pages 130–137, 2017. DOI: 10.1145/3109859.3109896 126, 127

[133] N. Quadrianto and C. H. Lampert. Learning multi-view neighborhood preserving projections. In *ICML*, pages 425–432, 2011. 93

[134] G. A. Ramirez, T. Baltrušaitis, and L.-P. Morency. Modeling latent discriminative dynamic of multi-dimensional affective signals. In *International Conference on Affective Computing and Intelligent Interaction*, pages 396–406, Springer, 2011. DOI: 10.1007/978-3-642-24571-8_51 25

[135] M. Ravanelli, B. Elizalde, K. Ni, and G. Friedland. Audio concept classification with hierarchical deep neural networks. In *IEEE EUSIPCO*, pages 606–610, 2014. 110

[136] M. Redi, N. O'Hare, R. Schifanella, M. Trevisiol, and A. Jaimes. 6 seconds of sound and vision: Creativity in micro-videos. In *Proc. of the IEEE Conference on Computer Vision and Pattern Recognition*, pages 4272–4279, 2014. DOI: 10.1109/cvpr.2014.544 3

[137] S. Rendle, C. Freudenthaler, Z. Gantner, and L. Schmidt-Thieme. BPR: Bayesian personalized ranking from implicit feedback. In *Proc. of the AUAI Conference on Uncertainty in Artificial Intelligence*, pages 452–461, 2009. DOI: 10.1142/s0218001416590011 126, 134

[138] S. D. Roy, T. Mei, W. Zeng, and S. Li. Towards cross-domain learning for social video popularity prediction. *IEEE Transactions on Multimedia*, 15(6):1255–1267, 2013. DOI: 10.1109/tmm.2013.2265079 24

[139] J. Rupnik and J. Shawe-Taylor. Multi-view canonical correlation analysis. In *Proc. of Conference on Data Mining and Data Warehouses*, pages 1–4, 2010. 39

[140] M. A. Saad, A. C. Bovik, and C. Charrier. Blind prediction of natural video quality. *IEEE Transactions on Image Processing*, 23(3):1352–1365, 2014. DOI: 10.1109/tip.2014.2299154 22

[141] S. Sadanand and J. J. Corso. Action bank: A high-level representation of activity in video. In *CVPR*, pages 1234–1241, 2012. DOI: 10.1109/cvpr.2012.6247806 109

[142] S. Sano, T. Yamasaki, and K. Aizawa. Degree of loop assessment in microvideo. In *IEEE International Conference on Image Processing*, pages 5182–5186, 2014. DOI: 10.1109/icip.2014.7026049 3

[143] G. Schindler, M. Brown, and R. Szeliski. City-scale location recognition. In *IEEE CVPR*, pages 1–7, 2007. DOI: 10.1109/cvpr.2007.383150 61

[144] E. Shutova, D. Kiela, and J. Maillard. Black holes and white rabbits: Metaphor identification with visual features. In *NAACL*, pages 160–170, 2016. DOI: 10.18653/v1/n16-1020 25

[145] V. Sindhwani, P. Niyogi, and M. Belkin. A co-regularization approach to semi-supervised learning with multiple views. In *Proc. of the International Conference on Machine Learning*, pages 74–79, ACM, 2005. 26

[146] A. W. M. Smeulders. Early vs. late fusion in semantic video analysis. In *ACM MM*, pages 399–402, 2005. DOI: 10.1145/1101149.1101236 25

[147] A. J. Smola and B. Schölkopf. A tutorial on support vector regression. *Statistics and Computing*, 14(3):199–222, 2004. DOI: 10.1023/b:stco.0000035301.49549.88 54

[148] C. G. Snoek, M. Worring, and A. W. Smeulders. Early vs. late fusion in semantic video analysis. In *Proc. of the ACM International Conference on Multimedia*, pages 399–402, 2005. DOI: 10.1145/1101149.1101236 59
DOI: 10.1145/2766462.2767726

[149] X. Song, L. Nie, L. Zhang, M. Akbari, and T.-S. Chua. Multiple social network learning and its application in volunteerism tendency prediction. In *Proc. of ACM SIGIR Conference on Research and Development in Information Retrieval*, pages 213–222, 2015. DOI: 10.1145/2766462.2767726 20, 34, 54, 55, 111

[150] X. Song, L. Nie, L. Zhang, M. Liu, and T.-S. Chua. Interest inference via structure-constrained multi-source multi-task learning. In *Proc. of the International Joint Conference on Artificial Intelligence*, pages 2371–2377, AAAI Press, 2015. 26, 67

[151] G. Szabo and B. A. Huberman. Predicting the popularity of online content. *Communications of the ACM*, 53(8):80–88, 2010. DOI: 10.2139/ssrn.1295610 42

[152] J. Tang and K. Wang. Personalized top-n sequential recommendation via convolutional sequence embedding. In *Proc. of the ACM International Conference on Web Search and Data Mining*, pages 565–573, 2018. DOI: 10.1145/3159652.3159656 126, 127

[153] K. Tang, M. Paluri, L. Fei-Fei, R. Fergus, and L. Bourdev. Improving image classification with location context. In *IEEE CVPR*, pages 1008–1016, 2015. DOI: 10.1109/iccv.2015.121 61

[154] G.-B. Huang, Q.-Y. Zhu, and C.-K. Siew. Extreme learning machine: Theory and applications. *Neurocomputing*, 70(1):489–501, 2006. DOI: 10.1016/j.neucom.2005.12.126 55

[155] L. C. Totti, F. A. Costa, S. Avila, E. Valle, W. Meira, and V. Almeida. The impact of visual attributes on online image diffusion. In *Proc. of ACM Web Science Conference*, pages 42–51, 2014. DOI: 10.1145/2615569.2615700 24

[156] T. Trzcinski and P. Rokita. Predicting popularity of online videos using support vector regression. *ArXiv Preprint ArXiv:1510.06223*, 2015. 24, 42

[157] T. X. Tuan and T. M. Phuong. 3D convolutional networks for session-based recommendation with content features. In *Proc. of ACM International Conference on Recommender Systems*, pages 138–146, 2017. DOI: 10.1145/3109859.3109900 126, 127

[158] M. Vasconcelos, J. M. Almeida, and M. A. Gonçalves. Predicting the popularity of micro-reviews: A foursquare case study. *Information Sciences*, 325:355–374, 2015. DOI: 10.1016/j.ins.2015.07.001 24

[159] D. Wang, S. C. H. Hoi, Y. He, J. Zhu, T. Mei, and J. Luo. Retrieval-based face annotation by weak label regularized local coordinate coding. *TPAMI*, 36(3):1–14, 2013. DOI: 10.1145/2072298.2072345 97

[160] D. Wang, X. Zhang, M. Fan, and X. Ye. Semi-supervised dictionary learning via structural sparse preserving. In *National Conference of the American Association for Artificial Intelligence*, pages 2137–2144, 2016. 62

[161] K. Wang, R. He, L. Wang, W. Wang, and T. Tan. Joint feature selection and subspace learning for cross-modal retrieval. *TPAMI*, 38(10):2010–2024, 2016. DOI: 10.1109/tpami.2015.2505311 26

[162] M. Wang, X. Hua, R. Hong, J. Tang, G. Qi, and Y. Song. Unified video annotation via multigraph learning. *IEEE Transactions on Circuits and Systems for Video Technology*, 19(5):733–746, 2009. DOI: 10.1109/tcsvt.2009.2017400 19

[163] M. Wang, H. Li, D. Tao, K. Lu, and X. Wu. Multimodal graph-based reranking for web image search. *IEEE Transactions on Image Processing*, 21(11):4649–4661, 2012. DOI: 10.1109/tip.2012.2207397 19

[164] M. Wang, X. Liu, and X. Wu. Visual classification by ℓ_1-hypergraph modeling. *IEEE Transactions on Knowledge and Data Engineering*, 27(9):2564–2574, 2015. DOI: 10.1109/TKDE.2015.2415497 30

[165] S. Wang, L. Zhang, Y. Liang, and Q. Pan. Semi-coupled dictionary learning with applications to image super-resolution and photo-sketch synthesis. In *IEEE Conference on Computer Vision and Pattern Recognition*, pages 2216–2223, 2012. DOI: 10.1109/cvpr.2012.6247930 62

[166] X. Wang, L. Nie, X. Song, D. Zhang, and T. S. Chua. Unifying virtual and physical worlds: Learning toward local and global consistency. *TOIS*, 36(1):1–26, 2017. DOI: 10.1145/3052774 92

[167] Y. Wang, S. Rawat, and F. Metze. Exploring audio semantic concepts for event-based video retrieval. In *IEEE ICASSP*, pages 1360–1364, 2014. DOI: 10.1109/icassp.2014.6853819 110

[168] M. White, Y. Yu, X. Zhang, and D. Schuurmans. Convex multi-view subspace learning. In *NIPS*, pages 1673–1681, 2012. 63

[169] S. Wold, K. Esbensen, and P. Geladi. Principal component analysis. *Chemometrics and Intelligent Laboratory Systems*, 2(1–3):37–52, 1987. DOI: 10.1016/0169-7439(87)80084-9 16

[170] B. Wu and H. Shen. Analyzing and predicting news popularity on twitter. *International Journal of Information Management*, 35(6):702–711, 2015. DOI: 10.1016/j.ijinfomgt.2015.07.003 24

[171] B. Wu, E. Zhong, A. Horner, and Q. Yang. Music emotion recognition by multi-label multi-layer multi-instance multi-view learning. In *Proc. of the ACM Multimedia Conference*, pages 117–126, ACM, 2014. DOI: 10.1145/2647868.2654904 21, 23

[172] F. Wu, X. Lu, Z. Zhang, S. Yan, Y. Rui, and Y. Zhuang. Cross-media semantic representation via bi-directional learning to rank. In *Proc. of the ACM International Conference on Multimedia*, pages 877–886, 2013. DOI: 10.1145/2502081.2502097 62

[173] J. Wu, Y. Zhou, D. M. Chiu, and Z. Zhu. Modeling dynamics of online video popularity. *IEEE Transactions on Multimedia*, 18(9):1882–1895, 2016. DOI: 10.1109/iwqos.2015.7404724 24

[174] Y. Wu and T. S. Huang. Self-supervised learning for visual tracking and recognition of human hand. In *National Conference of the American Association for Artificial Intelligence*, pages 243–248, 2000. 62

[175] Z. Wu, Y. Jiang, J. Wang, J. Pu, and X. Xue. Exploring inter-feature and inter-class relationships with deep neural networks for video classification. In *Proc. of the ACM Multimedia Conference*, pages 167–176, 2014. DOI: 10.1145/2647868.2654931 23

[176] R. Xia, Y. Pan, L. Du, and J. Yin. Robust multi-view spectral clustering via low-rank and sparse decomposition. In *Proc. of AAAI Conference on Artificial Intelligence*, pages 2149–2155, 2014. 27

[177] S. Xiang, Y. Zhu, X. Shen, and J. Ye. Optimal exact least squares rank minimization. In *Proc. ACM SIGKDD International Conference on Knowledge Discovery and Data Mining*, pages 480–488, 2012. DOI: 10.1145/2339530.2339609 27

[178] Y. Yan, Y. Yang, H. Shen, D. Meng, G. Liu, A. G. Hauptmann, and N. Sebe. Complex event detection via event oriented dictionary learning. In *National Conference of the American Association for Artificial Intelligence*, pages 3841–3847, 2015. 62

[179] J. Yang, J. Wright, T. S. Huang, and Y. Ma. Image super-resolution via sparse representation. *IEEE Transactions on Image Processing*, 19(11):2861–2873, 2010. DOI: 10.1109/tip.2010.2050625 62

[180] M. Yang, W. Liu, W. Luo, and L. Shen. Analysis-synthesis dictionary learning for universality-particularity representation based classification. In *National Conference of the American Association for Artificial Intelligence*, pages 2251–2257, 2016. 62

[181] Y. Yang, J. Song, Z. Huang, and Z. Ma. Multi-feature fusion via hierarchical regression for multimedia analysis. *IEEE Transactions on Multimedia*, 15(3):572–581, 2013. DOI: 10.1109/tmm.2012.2234731 54

[182] Z. Yang, W. W. Cohen, and R. Salakhutdinov. Revisiting semi-supervised learning with graph embeddings. In *ICML*, pages 40–48, 2016. 114

[183] M. Ye, P. Yin, and W.-C. Lee. Location recommendation for location-based social networks. In *ACM AGIS*, pages 458–461, 2010. DOI: 10.1145/1869790.1869861 60

[184] Y. Ying and C. Campbell. Rademacher chaos complexities for learning the kernel problem. *Neural Computation*, vol. 22(11), pages 2858–2886, 2010. 26

[185] J. Yosinski, J. Clune, Y. Bengio, and H. Lipson. How transferable are features in deep neural networks? In *NIPS*, pages 3320–3328, 2014. 118

[186] S. Yu, B. Krishnapuram, R. Rosales, and R. B. Rao. Bayesian co-training. *Journal of Machine Learning Research*, 12(3):2649–2680, 2011. 25

[187] D. Zhai, H. Chang, S. Shan, X. Chen, and W. Gao. Multiview metric learning with global consistency and local smoothness. *TIST*, 3(3):1–22, 2012. DOI: 10.1145/2168752.2168767 26

[188] H. Zhang, X. Shang, W. Yang, H. Xu, H. Luan, and T. Chua. Online collaborative learning for open-vocabulary visual classifiers. In *Proc. of the IEEE Conference on Computer Vision and Pattern Recognition*, 2016. DOI: 10.1109/cvpr.2016.307 22

[189] H. Zhang, M. Wang, R. Hong, and T. Chua. Play and rewind: Optimizing binary representations of videos by self-supervised temporal hashing. In *Proc. of the ACM Multimedia Conference*, 2016. DOI: 10.1145/2964284.2964308 19
DOI: 10.1145/2339530.2339617

[190] J. Zhang and J. Huan. Inductive multi-task learning with multiple view data. In *Proc. of ACM International Conference on Knowledge Discovery and Data Mining*, pages 543–551, 2012. DOI: 10.1145/2339530.2339617 34, 54, 62, 68

[191] J. Zhang, L. Nie, X. Wang, X. He, X. Huang, and T.-S. Chua. Shorter-is-better: Venue category estimation from micro-video. In *Proc. of ACM International Conference on Multimedia, (MM)*, pages 1415–1424, 2016. DOI: 10.1145/2964284.2964307 109, 110

[192] J. Zhang, L. Nie, X. Wang, X. He, X. Huang, and T.-S. Chua. Shorter-is-better: Venue category estimation from micro-video. In *Proc. of the ACM International Conference on Multimedia*, pages 1–10, 2016. DOI: 10.1145/2964284.2964307 59, 82, 85, 90, 101
DOI: 10.1145/2964284.2964307

[193] Q. Zhang and B. Li. Discriminative K-SVD for dictionary learning in face recognition. In *IEEE Conference on Computer Vision and Pattern Recognition*, pages 2691–2698, 2010. DOI: 10.1109/cvpr.2010.5539989 62

[194] W. Zhang, C. Liu, Z. Wang, G. Li, Q. Huang, and W. Gao. Web video thumbnail recommendation with content-aware analysis and query-sensitive matching. *Multimedia Tools and Applications*, 73(1):547–571, 2014. DOI: 10.1007/s11042-013-1607-5 145

[195] X. Zhang, F. Sun, G. Liu, and Y. Ma. Fast low-rank subspace segmentation. *IEEE Transactions on Knowledge and Data Engineering*, 26(5):1293–1297, 2014. DOI: 10.1109/tkde.2013.114 39

[196] Y. Zhang, Z. Jiang, and L. S. Davis. Learning structured low-rank representations for image classification. In *Proc. of IEEE Conference on Computer Vision and Pattern Recognition*, pages 676–683, 2013. DOI: 10.1109/cvpr.2013.93 27

[197] Z. Zhang, F. Li, M. Zhao, L. Zhang, and S. Yan. Joint low-rank and sparse principal feature coding for enhanced robust representation and visual classification. *IEEE Transactions on Image Processing*, 25(6):2429–2443, 2016. DOI: 10.1109/tip.2016.2547180 26, 39

[198] Z. Zhang, F. Li, M. Zhao, L. Zhang, and S. Yan. Robust neighborhood preserving projection by nuclear/l2, 1-norm regularization for image feature extraction. *IEEE Transactions on Image Processing*, 26(4):1607–1622, 2017. DOI: 10.1109/tip.2017.2654163 26

[199] Z. Zhang, S. Yan, and M. Zhao. Similarity preserving low-rank representation for enhanced data representation and effective subspace learning. *Neural Networks*, 53:81–94, 2014. DOI: 10.1016/j.neunet.2014.01.001

[200] Z. Zhang, M. Zhao, F. Li, L. Zhang, and S. Yan. Robust alternating low-rank representation by joint lp-and l2, p-norm minimization. *Neural Networks*, 96:55–70, 2017. DOI: 10.1016/j.neunet.2017.08.001 26

[201] X. Zhao, G. Li, M. Wang, J. Yuan, Z.-J. Zha, Z. Li, and T.-S. Chua. Integrating rich information for video recommendation with multi-task rank aggregation. In *Proc. of the ACM International Conference on Multimedia*, pages 1521–1524, 2011. DOI: 10.1145/2072298.2072055 125, 126, 127

[202] S. Zheng, X. Cai, C. H. Ding, F. Nie, and H. Huang. A closed form solution to multi-view low-rank regression. In *Proc. of AAAI Conference on Artificial Intelligence*, pages 1973–1979, 2015. 27

[203] C. Zhou, J. Bai, J. Song, X. Liu, Z. Zhao, X. Chen, and J. Gao. Atrank: An attention-based user behavior modeling framework for recommendation. In *Proc. of the AAAI Conference on Artificial Intelligence*, 2018. 126, 127, 134

[204] J. Zhou, J. Chen, and J. Ye. Malsar: Multi-task learning via structural regularization. In *Arizona State University*, pages 1–50, 2011. 64

[205] P. Zhou, Z. Lin, and C. Zhang. Integrated low-rank-based discriminative feature learning for recognition. *IEEE Transactions on Neural Networks and Learning Systems*, 27(5):1080–1093, 2016. DOI: 10.1109/tnnls.2015.2436951 27

[206] X. Zhou, L. Chen, Y. Zhang, L. Cao, G. Huang, and C. Wang. Online video recommendation in sharing community. In *Proc. of the ACM SIGMOD International Conference on Management of Data*, pages 1645–1656, 2015. DOI: 10.1145/2723372.2749444 125, 126

[207] X. Zhou, M. Zhu, S. Leonardos, and K. Daniilidis. Sparse representation for 3D shape estimation: A convex relaxation approach. *TPAMI*, PP(99):1–14, 2015. DOI: 10.1109/tpami.2016.2605097 97

[208] Z. H. Zhou and M. Li. Semi-supervised regression with co-training. In *IJCAI*, pages 908–913, 2005. 25

[209] Q. Zhu, M.-L. Shyu, and H. Wang. Videotopic: Content-based video recommendation using a topic model. In *Proc. of the IEEE International Symposium on Multimedia*, pages 219–222, 2013. DOI: 10.1109/ism.2013.41 125, 126

[210] J. Zhuang, T. Mei, S. C. Hoi, X.-S. Hua, and Y. Zhang. Community discovery from social media by low-rank matrix recovery. *ACM Transactions on Intelligent Systems and Technology*, 5(4):67:1–19, 2015. DOI: 10.1145/2668110 26

Authors' Biographies

LIQIANG NIE

Liqiang Nie is currently a professor with the School of Computer Science and Technology, Shandong University. In addition, he is the adjunct dean with the Shandong AI institute. He received his B.Eng. and Ph.D. from Xi'an Jiaotong University in 2009 and the National University of Singapore (NUS) in 2013, respectively. After his Ph.D., Dr. Nie continued his research in NUS as a research fellow for three and half years. His research interests lie primarily in multimedia computing and information retrieval. Dr. Nie has authored and/or coauthored more than 100 papers for SIGIR, ACM MM, TOIS, and TIP, received more than 4,800 Google Scholar citations. He is an AE of Information Science, and an area chair of ACM MM 2018/2019.

MENG LIU

Meng Liu is currently a Ph.D. student with the School of Computing Science and Technology, Shandong University. She received an M.S. in computational mathematics from Dalian University of Technology, China in 2016. Her research interests are multimedia computing and information retrieval. Various parts of her work have been published in top forums and journals, such as SIGIR, MM, and IEEE TIP. She has served as reviewers and subreviewers for various conferences and journals, such as MMM, MM, PCM, JVCI, and INS.

XUEMENG SONG

Xuemeng Song received a B.E. from the University of Science and Technology of China in 2012, and a Ph.D. from the School of Computing, National University of Singapore in 2016. She is currently an assistant professor of Shandong University, Jinan, China. Her research interests include the information retrieval and social network analysis. She has published several papers in the top venues, such as ACM SIGIR, MM, and TOIS. In addition, she has served as a reviewer for many top conferences and journals.